Chemistry

by Ian Guch

A member of Penguin Group (USA) Inc.

ALPHA BOOKS

Published by the Penguin Group

Penguin Group (USA) Inc., 375 Hudson Street, New York, New York 10014, U.S.A.

Penguin Group (Canada), 10 Alcorn Avenue, Toronto, Ontario, Canada M4V 3B2 (a division of Pearson Penguin Canada Inc.)

Penguin Books Ltd, 80 Strand, London WC2R 0RL, England

Penguin Ireland, 25 St Stephen's Green, Dublin 2, Ireland (a division of Penguin Books Ltd)

Penguin Group (Australia), 250 Camberwell Road, Camberwell, Victoria 3124, Australia (a division of Pearson Australia Group Pty Ltd)

Penguin Books India Pvt Ltd, 11 Community Centre, Panchsheel Park, New Delhi—110 017, India

Penguin Group (NZ), cnr Airborne and Rosedale Roads, Albany, Auckland 1310, New Zealand (a division of Pearson New Zealand Ltd)

Penguin Books (South Africa) (Pty) Ltd, 24 Sturdee Avenue, Rosebank, Johannesburg 2196, South Africa

Penguin Books Ltd, Registered Offices: 80 Strand, London WC2R 0RL, England

International Standard Book Number: 1-59257-350-9
Library of Congress Catalog Card Number: 2004115799

07 06 05 8 7 6 5 4 3 2 1

Interpretation of the printing code: The rightmost number of the first series of numbers is the year of the book's printing; the rightmost number of the second series of numbers is the number of the book's printing. For example, a printing code of 05-1 shows that the first printing occurred in 2005.

Printed in the United States of America

This book is dedicated to my wife, Ingrid, who makes every day sunny and bright.

Contents

Introduction

You need help with chemistry. Either you're taking a class and want a review that will help you go through the basic concepts, or you're somebody who's interested in chemistry and wants to know what it's all about. Either way, you're hoping that this book will give you everything you need to know in a few hundred pages, preferably in a form that doesn't use too many scary chemistry terms.

Well, I have good news and bad news. The good news is that this book does, in fact, cover introductory chemistry in terms that shouldn't be too terrifying. By the time you're done with this book, you'll have reviewed the vast majority of the information that's covered in first-year chemistry courses, and have a pretty good idea about how to understand the rest.

The bad news is that it doesn't cover everything there is to know about chemistry. As you've probably already realized, chemistry is a *really* big subject and covers a huge amount of material. This book doesn't pretend to tell you everything, because frankly, there's no book in the universe that can do that. What this book should do, however, is give you the background information you'll need if you're going to attack some of the more advanced topics in modern chemistry.

In addition to just spouting out pure chemistry knowledge, I've also attempted to bring you this information in a way that shows you how fun chemistry really is. Sure, not all of the topics are endless thrill rides, but I think you'll find that if you pause every so often and think about the uses of this stuff in the real world, you'll realize that chemistry is neither as hard nor as irrelevant as you might have thought. In fact, chemistry is a huge amount of fun, and is one of the primary tools that scientists use to improve the world around us.

I won't promise that you'll end up with a Nobel Prize in chemistry, or even that this book will be your first step toward curing cancer. What I will promise is that if you really understand the material in these pages, you'll have all of the background chemistry information you need to understand the chemistry around you (and pass general chemistry).

How to Use This Book

Because this is a review of general chemistry, you'll find that each chapter covers a lot of ground, building on the ones before it. If you have a pretty good grasp of the more basic concepts, you can probably skip straight through to the chapter on whatever concept is bothering you. However, if you need a general review of the entire subject, I strongly suggest that you start from the beginning and move through the book one chapter at a time. While you'll undoubtedly see things that you're already familiar with, I think you'll find that each topic really does make sense, if you understand the ones before it.

I also strongly encourage you to solve sample problems on your own as you move through this book. Although I haven't included unsolved sample problems for you to do on your own due to space constraints, the only way for you to really understand what you're studying is to try it on your own and see if you get the right answer. As you travel through this book, find sample problems in a chemistry textbook or online to see if you really understand what you're reading. If you get them right, you're on the right track!

Helpful Reminders and Random Thoughts

As you'll quickly realize, there are a lot of terms and important ideas to remember when studying chemistry. Because it's no fun to try and find the most important of these terms and ideas by wading through the text, I've included them as sidebars:

The Mole Says

These boxes highlight the most important things you'll need to know from each chapter. If you understand these concepts, you'll be in good shape.

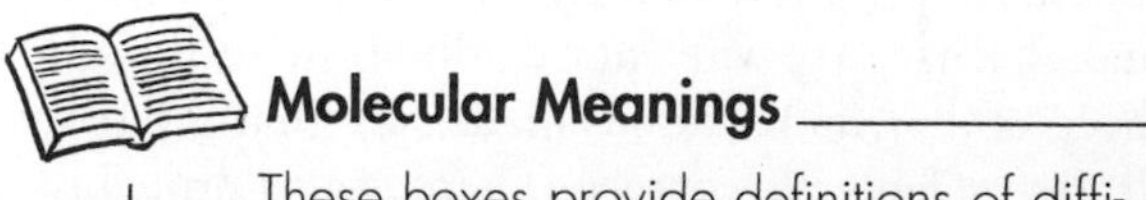

Molecular Meanings

These boxes provide definitions of difficult or unfamiliar terms.

Bad Reactions

These represent common mistakes that beginning chemistry students make, and give suggestions for how to avoid them.

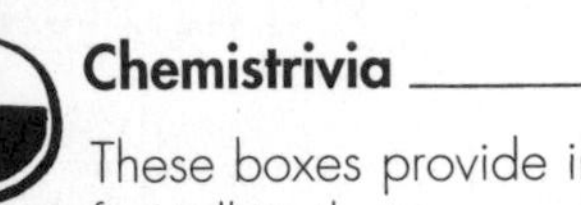

Chemistrivia

These boxes provide interesting and fun tidbits that can make chemistry a little more fun.

Acknowledgments

I'd like to take a few paragraphs to thank the people who made this book possible.

First off, I'd like to thank Jessica Faust from BookEnds for convincing Alpha to let me write this book, and Michael Thomas and Michael Sanders from Alpha for agreeing to it and answering my many questions.

I'd also like to thank the folks who have been involved in turning this book from a vast matrix of run-on sentences and mistakes into something that actually makes sense. My wife Ingrid edited the terrible first draft, and Orna Kutai has made sure that the second draft was factually correct. There is also probably a huge army of proofreaders at Alpha who have read this, but I don't know their names. Thanks, whoever you are!

There are also other people for whom special thanks is deserved, even if they didn't participate directly in the making of this book. Thanks to Marcello DiMare and Bruce Rickborn for giving me a love and understanding of chemistry. Thanks to Nancy Levinger, who convinced me that I make a much better writer and teacher than chemist. Thanks to those who have helped me become a better teacher, including Joy McManus, Jan Warner, and Anita Scovanner-Ramsey. Thanks to my friends who put up with me all these years, including Laurie and Kat Wallace, Donna Delano, Pam Meier, Kevin Whelan, Erik Luther, Rob Keil and Linda Williamson, Brad Luther and Corinna Cincotta, Marni Frank, Shaun Rafferty, Bill Bradshaw, Meikka Cutlip, Scott Pasch, the Pheer folk, Diane Cunningham, Fr. Jim Shea, Fr. Jim Seebak, Jeff Schooner, and my neighbors who don't complain about the loud punk rock. Thanks to my family, including my parents, Matt, Cindy, Owen, Zachary, my grandparents, and Ilga.

Thanks most of all to my wife, Ingrid. She's the most wonderful person I know, and I'm unbelievably lucky to have her in my life.

And, lastly, thanks to you, the reader. After all, they wouldn't pay me to write a book if you didn't pay to buy it! Keep up the good work, and buy copies of this book for all your friends!

Trademarks

Chapter 1

All About the Atom

In This Chapter

- The historical development of the atom
- The quantum mechanical model of the atom
- Isotopes and subatomic particles

Atoms are very, very small. That's probably something you already know and understand fairly well. However, what's *really* up with atoms? What rules govern how they behave?

I'm glad you asked! Let's take a look …

What's an Atom?

Atoms are the smallest chunks of an element that have the same properties as larger chunks of that element. If you take a piece of iron and break it in half over and over again, you'll eventually end up with a very tiny piece that can't be broken apart any more. That's an atom.

Molecular Meanings

An **atom** is the smallest chunk of an element that has the same properties as larger chunks of that element.

Older Atomic Theories—What We Used to Think

Although it's not that uncommon nowadays for scientists to use scary-looking equipment to look at atom-sized pieces of matter, that wasn't always the case. As a result, even the brightest minds of the past had some pretty odd ideas about what atoms are like. Let's check them out.

It's All Greek to Me

The first people we know of who started to think about the nature of matter were the ancient Greeks. They suggested that when you try to break matter apart, you eventually end up with very tiny indivisible particles called atoms. Because nobody had any way to test whether or not this was right, this theory reigned for the next two millennia.

Chemistrivia

The ancient Greeks also thought the only four elements were earth, air, fire, and water, and that all matter was made of some combination of these elements. Presumably, chili peppers would be mostly fire.

Some Interesting Observations

For a very long time, the Greek model of the atom was unchallenged. However, in the late eighteenth century, new information changed all that. Let's summarize some of the most interesting points:

- The **law of conservation of mass** states that the weight of the products of a chemical reaction is the same as the weight of the reactants that you started with. In other words, if you make a pizza, the final weight of the pizza will be equal to the weight of the ingredients.
- The **law of definite composition** states that the ratio of the elements in a chemical compound always stays the same. That means that no matter how you make a molecule of water, the formula will always be H_2O.
- The **law of multiple proportions** states that when two elements combine to form more than one chemical compound, the ratio of the masses of one element that combines with a fixed mass of the other element can

be expressed as a ratio of small, whole numbers. For example, two compounds can be made from the combination of hydrogen and oxygen: water (H_2O) and hydrogen peroxide (H_2O_2). When you divide the mass of the oxygen in hydrogen peroxide with the mass of the oxygen in water, you get a ratio of 2:1. This works for other compounds, too, though the ratio may be 3:1 or 4:3 or whatever.

Dalton's Laws

In 1808, John Dalton used these new discoveries to come up with a series of rules that describe how atoms behave, creatively referred to as **Dalton's Laws:**

- All matter is made of tiny, indestructible particles called atoms.
- All atoms of a given element have identical chemical and physical properties.
- Atoms of different elements have different sets of chemical and physical properties.
- Atoms are neither created nor destroyed, and they obey the law of conservation of mass.
- Atoms of different elements form compounds in whole-number ratios. For example, water is H_2O, not $H_{2.1}O_{1.3}$.

Several of these rules are now known to have exceptions, though it was impossible for Dalton to know this based on the information he had. We now know that atoms can be broken apart through nuclear reactions. Atoms of a given element may also not have identical properties if they are isotopes (more about that later in this chapter).

J. J. Thomson—A Man and His Dessert

In 1897, British physicist J. J. Thomson performed an experiment in which voltage was applied across two wires (called **electrodes**) in a vacuum tube. As a result of this applied voltage, a glowing beam of particles was observed to travel from the **cathode** (the negative electrode) to the anode (the positive electrode). Because these particles originated at the cathode, they were called, creatively enough, **cathode rays.**

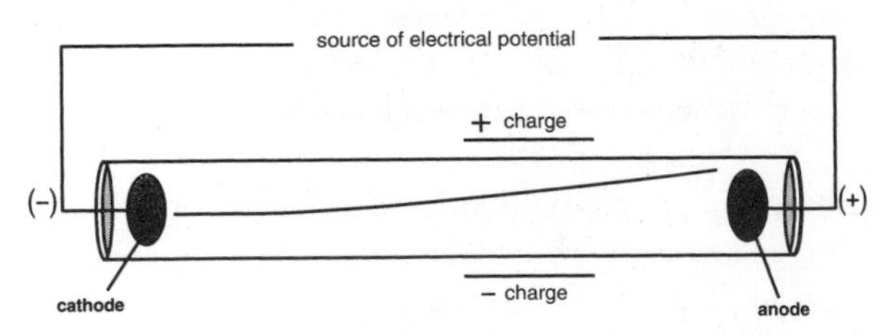

Thomson found that cathode rays moved away from negative charge and toward positive charge.

When Thomson studied how these particles behaved near magnets, he found that they tended to move toward the positive poles and away from the negative poles. From this, he determined that the rays were made of very tiny particles with negative charge. Because he saw no corresponding positive charges moving from one side of the tube to the other, he concluded that the positive charge in an atom is relatively heavy and immobile. From this knowledge he came up with a model of the atom he referred to as the **"plum pudding" model.**

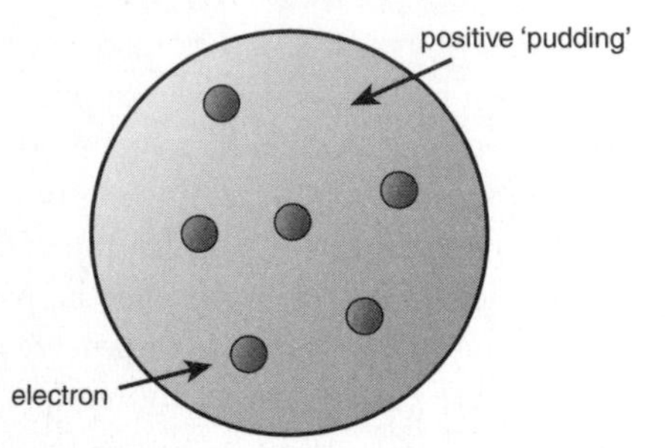

Named after his dessert, Thomson's plum pudding model portrays the atom as a big ball of positive charge that contains small particles of negative charge. In this picture, the dough represents the positive charge and the chips represent the electrons.

Because practically nobody knows what plum pudding is anymore, think of a chocolate chip cookie instead. Thomson believed that the dough was a big blob of positive charge into which the chips (electrons) were embedded. Rumor has it that Thomson was working on an "oatmeal raisin" model of the atom when he passed away, but unfortunately this work was lost.

Rutherford's Alpha Particles

One day in 1907, Ernst Rutherford was playing with radioactive particles in his lab. Specifically, he was shooting alpha particles at a very thin piece of gold foil to see what would happen. Alpha particles, which we'll discuss in more detail in Chapter 11, consist of helium nuclei with a +2 charge.

What he found was that most of the particles passed right through the foil, some were slightly deflected, and others bounced right back at the source. From this discovery Rutherford suggested that the atom is mostly empty space, which would explain why most particles passed through the foil without changing direction. He explained the various deflections of the other particles by theorizing that all of the positive charge in the atom is concentrated into one small area (now called the **nucleus**). The severity of the deflection of the positively charged alpha particles depended on how close they traveled to the positively charged nucleus. Furthermore, Rutherford theorized that the electrons in an atom float around the nucleus.

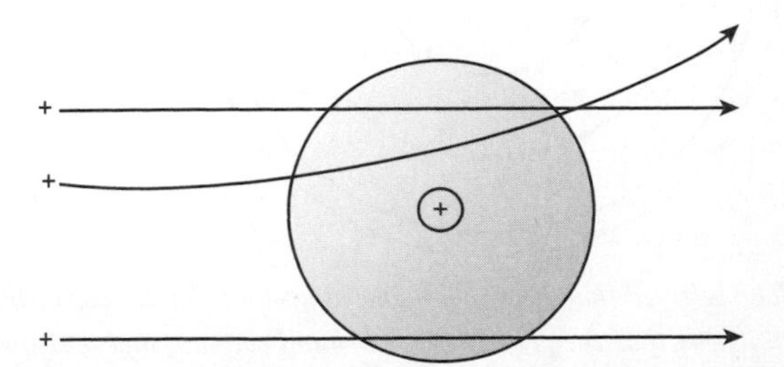

Rutherford believed that when positively charged alpha particles passed near the positively charged nucleus, the resulting strong repulsion caused them to be deflected at extreme angles.

Bohr: What Is It Good For?

Neils Bohr wasn't the kind of guy who liked to play with cathode ray tubes or alpha particles. Instead, while the other scientists were playing with their cool equipment, Bohr was busy thinking about hydrogen.

Specifically, Bohr wondered why, when you add energy to hydrogen, only certain energies of light are given off (these energies correspond to different colors of light). After all, there's nothing about any of the earlier models of the atom suggesting that atoms should give off light at all, much less light with specific colors. After much thought, Bohr came up with his **planetary model** of the atom.

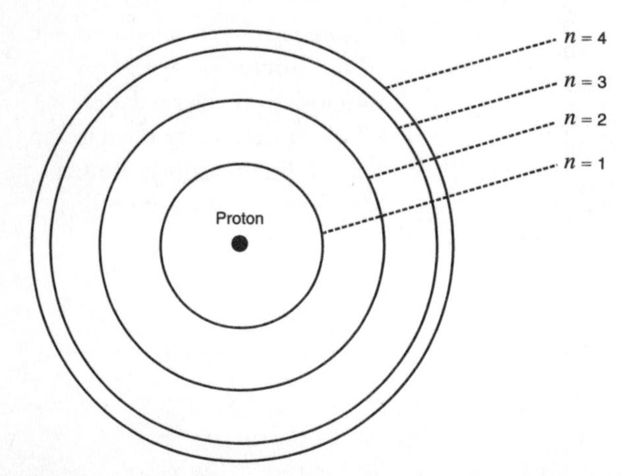

Bohr believed that electrons traveled around the nucleus in the same way that the planets move around the sun, and that their distances from the nucleus were related to their energy.

In his planetary model, Bohr suggested that electrons travel around the nucleus in circular orbits, just as the planets revolve around the sun. The paths that these electrons traveled were called orbitals, and the further the orbitals were from the nucleus, the more energy the electrons within them had. For convenience, the orbitals were denoted by the variable n, with the nearest orbital to the nucleus being equal to $n = 1$, the second nearest being $n = 2$, and so on.

This fit nicely with his observation of the energies of light given off by hydrogen. Bohr theorized that when energy is added to an atom, electrons jump from their original orbital (called the ground state) to an orbital with higher energy (called the excited state). When the electrons eventually fall back down to their ground state, the energy they initially absorbed is given off all at once as light. Because every element gives off a unique pattern of colors (called a **line spectrum**), we can use these characteristic energies to tell elements apart from each other. The method of identifying substances from these energies is called **spectroscopy.**

Quantum Mechanics

Unfortunately for Bohr, this model didn't predict the colors in the spectrum for any element but hydrogen. However, since it did work for hydrogen, that got other people thinking that maybe this model was at least partially right. After a lot of work, the **quantum model** of the atom was born.

The Quantum Model of the Atom

The quantum model made Bohr's model more complex to better explain the line spectra of elements other than hydrogen. By the addition of different variables, the equations for orbitals stopped being perfect circles and turned into three-dimensional shapes. Furthermore, it was no longer possible to tell exactly where an electron would be found within each orbital—instead, we could only predict how likely it would be to find an electron in a particular location.

To make this somewhat more understandable, think of an orbital as being like a baseball field and an electron as being like a baseball. Imagine that, for whatever odd reason, we blindfold a very poor pitcher and walk him to the pitcher's mound. For good measure, we'll give him a spin so he won't know what direction he's facing. Can you predict exactly where the ball will go?

Clearly, you can't. However, you can make some reasonable guesses about where the ball might end up. For example, you know that a weak pitcher will probably not throw the ball very far, so it will end up somewhere in the vicinity of the pitcher's mound. Because the pitcher won't know what direction he's throwing, you can also imagine that the ball has an equal chance of going in any direction. In fact, if he throws the ball a hundred times, you can imagine that the balls will end up roughly in these locations shown in the following figure.

Although you may not know exactly where the ball will end up, you can make some guesses about it based on what you already know.

Orbitals are like this diagram. You can never tell exactly where an electron will be within an orbital, but the very nature of the orbital does give you some idea about the locations that are most likely.

Quantum Numbers

In quantum mechanics, four different variables are used to describe the locations of the electrons in an atom. These four variables are called, straightforwardly enough, **quantum numbers:**

- The **principal quantum number,** n, is used to describe the energy level of an electron. The allowed values for n are 1, 2, 3, … to infinity.
- The **angular momentum quantum number,** l, is used to determine the shape and type of the orbital. Possible values for l are 0, 1, 2, and so on up to $(n - 1)$. For example, if $n = 2$, the possible values for l are 0 and 1.

S-orbitals are defined as $l = 0$, p-orbitals are defined as $l = 1$, d-orbitals are defined as $l = 2$, and f-orbitals are defined as $l = 3$. Within each energy level, s-orbitals have the lowest energies, followed by p-orbitals, d-orbitals, and f-orbitals.

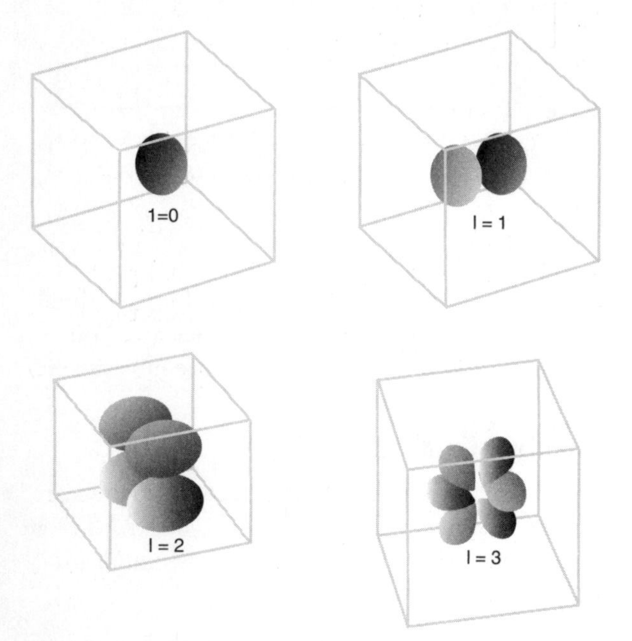

The value of l determines the shape and type of orbital being described.

The Mole Says

The magnetic quantum number explains why in every energy level there are a maximum of 1 s-orbital, 3 p-orbitals, 5 d-orbitals, and 7 f-orbitals. However, each type of orbital can hold a maximum of two electrons.

- The **magnetic quantum number,** m_l, determines the direction that the orbital points in space. Possible values for m_l include all the integers from $-l$ through $+l$. For example, if p-orbitals $l = 1$, the possible values of m_l are –1, 0, 1. Because of this, there are three possible directions that p-orbitals can point.

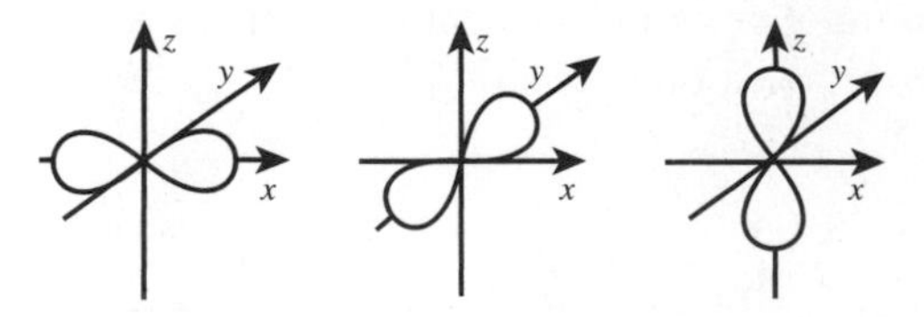

The three p-orbitals lie on the x-axis, the y-axis, and the z-axis.

- The **spin quantum number,** m_s, allows us to tell the two electrons in a single orbital apart. Possible values for m_s are +½ and –½.

Electron Configurations

Electrons fill up low-energy orbitals before moving to higher-energy orbitals, so we can write the ground state locations of all of the electrons in an atom using something called an **electron configuration.** Electron configurations usually contain many terms, each of which has this general format: ***n* (type of orbital)**$^{\text{number of electrons in that type of orbital}}$, where *n* refers to the principal quantum number described earlier. For example, carbon has the electron configuration $1s^2 2s^2 2p^2$, denoting that two electrons are present in the 1s orbital, two electrons are present in the 2s orbital, and two electrons are present in the 2p orbital. Generally, you'll find that the electron configuration for an element is exactly the same as the one for the element before it, except that the last term has changed. This occurs because of the **Aufbau principle,** which states that every element has the same number and placement of electrons as the one before it, plus one extra.

> **The Mole Says**
>
> Electron configurations typically show the ground state locations of the electrons in an atom.

Orbital Filling Diagrams and Hund's Rule

Because electrons all have negative charge, they don't like to be close together unless they absolutely have to. For example, let's consider the case of carbon, which has the electron configuration $1s^22s^22p^2$. Because there's only one s-orbital per energy level, each s-orbital has two electrons paired up in them. However, since there are three p-orbitals per energy level, the two p-electrons in this atom can occupy different orbitals, as shown:

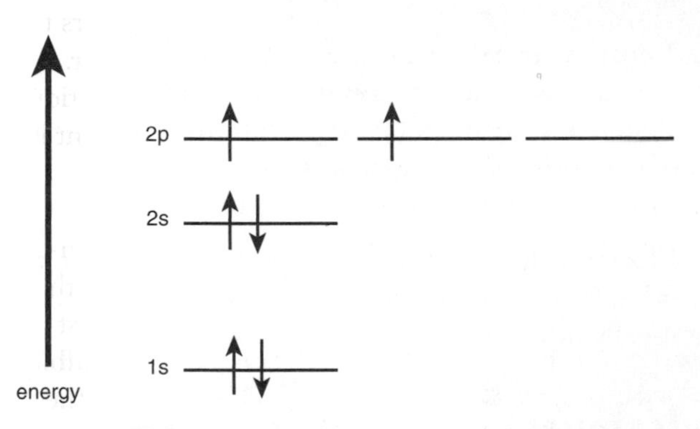

The orbital filling diagram for carbon.

The rule that electrons prefer to stay unpaired whenever possible in orbitals with equal energies is called **Hund's rule.**

The Parts of an Atom

The three fundamental particles that make up an atom are the protons and neutrons, located in the nucleus, and the electrons, located in the orbitals.

Protons each have a charge of +1 and are located in the nucleus (giving it the positive charge that

Rutherford noticed). Their mass is approximately 1 atomic mass unit, abbreviated as amu (1 amu weighs about 1.67×10^{-27} kg). The number of protons in an atom is referred to as the **atomic number**, and determines what element is present. For example, elements with one proton are always hydrogen, regardless of how many neutrons or electrons are present.

Neutrons each have no charge and are also located in the nucleus. Like protons, they weigh approximately 1 amu. The main function of neutrons is to keep all of the positively charged protons in the nucleus spaced out so the nucleus doesn't break apart.

Electrons have a charge of –1 and are located in the orbitals. They have a mass of $1/1836$ amu, which we usually just round to zero.

Isotopes

Isotopes are forms of an element that have different numbers of neutrons. This occurs because there are usually several numbers of neutrons that can keep the protons in the nucleus of an atom stable. Because these isotopes have different numbers of neutrons, they also have different masses.

To tell the difference between the different isotopes of an element, scientists have invented **A, Z, X nomenclature,** shown as ${}^{A}_{Z}X$. In this system, Z stands for the atomic number (the number of protons in the atom), A stands for the mass number of the isotope (the number of protons plus the number of neutrons), and X stands for the atomic symbol on the periodic table that denotes each element. For example, the form of carbon with seven neutrons is written as ${}^{13}_{6}C$, to show that it has six protons (all atoms of carbon have six protons!) and a mass of 13 amu (six protons + seven neutrons). Another common way to denote the isotopes of an element is to say the name of the element, followed by its mass. As a result, ${}^{13}_{6}C$ can also be properly referred to as "carbon-13."

Elements, Compounds, and Mixtures

In This Chapter

- Elements, compounds, and mixtures
- The organization of the periodic table
- Periodic trends

In Chapter 1, we learned what atoms were all about. Of course, chemistry is more complicated than that—if it weren't, we wouldn't need to know anything about chemical compounds and reactions. In this chapter, we're going to take the next step toward reactions by learning about elements, compounds, and mixtures.

Pure Substances

If somebody were to ask you whether you'd like to drink pure water or impure water, you'd probably answer pure water (and give them a funny look), because you'd like your cool beverage to be free of floating debris. Not surprisingly, that's what we chemists think about when we refer to something as being pure. To a chemist, a **pure substance** is one in which there's only one type of material present. Let's look at the two different types of pure substances:

- *Elements* are substances that contain only one type of atom. A full list of the known elements is shown in the periodic table, which we'll be discussing later in this chapter.
- *Compounds* are substances that contain only one type of chemical compound. Because of this, they always have a chemical name (e.g., sodium chloride) and chemical formula (e.g., NaCl). These substances can be broken apart by chemical means into the elements that make them up.

Molecular Meanings

Elements are pure substances containing only one type of atom, and **compounds** are pure substances containing only one type of chemical compound.

Mixtures

Mixtures are materials that contain more than one type of element or compound. Common examples of mixtures include air, jelly beans, and TV talk show host Jerry Springer. Anything, as long as it contains more than one type of element or compound, is a mixture.

- **Homogeneous mixtures** (also called solutions) are mixtures that are uniform in composition. Common examples of homogeneous mixtures include salt water, air, and stainless steel.
- **Heterogeneous mixtures** are mixtures that have an uneven composition. Let's consider my world-famous chili; your first spoonful might contain a bean and the second might contain beef. Other examples of heterogeneous mixtures include a chef salad, my nose, and punk rocker Lee Ving.
- **Colloids** are mixtures in which one type of particle is suspended in another without actually having been dissolved. It's not quite a solution because you can filter out the particles of suspended stuff, and it's not quite a heterogeneous mixture because it has a nearly uniform composition. Common examples of colloids include paint, mayonnaise, and smoke.

Let's Get Organized: The Periodic Table

If you're anything like me, you don't like to memorize things. Wouldn't it be nice if we could just use a cheat sheet to get through chemistry?

As it turns out, chemists already have a pretty good cheat sheet called the **periodic table.** What's even better is that even if you tell your instructor that you're using this cheat sheet, he or she won't be able to stop you!

Chemistrivia

The periodic table was invented in 1871 by Russian chemistry professor Dmitri Mendeleev. Though his periodic table was slightly different than our modern table, he was able to arrange the elements in a way that not only gave chemists a good "cheat sheet" but also allowed for the prediction of several previously undiscovered elements.

Metals, Nonmetals, and Metalloids

Most periodic tables have a staircase line that starts to the left of boron and moves downward and to the right. The elements to the left of this staircase are **metals,** which are generally shiny, hard, *malleable,* ductile, and conduct electricity. The elements to the right of this staircase (and hydrogen) are **nonmetals,** which are elements that are not shiny, malleable, *ductle*, and don't conduct electricity. The elements immediately surrounding this line (B, Si, Ge, As, Sb, Te, and Po) are **metalloids,** which have properties halfway between that of metals and nonmetals; some are shiny, they're semiconductors of electricity, and they are brittle.

Molecular Meanings

Materials that are **malleable** are able to be squished into thin sheets, and **ductile** materials are able to be pulled into wires. Put together, these terms roughly translate into "bendy."

Periodic Families

Elements in the same column of the periodic table are referred to as being **families** or **groups** (the terms are interchangeable). Because elements in the same family have similar electron configurations (see Chapter 1), they also have similar properties. Some of the most important families of the periodic table are shown below:

- Group 1 (except for hydrogen) is the **alkali metals.** Alkali metals are highly reactive elements that are very soft and have very low densities.
- Group 2 is the **alkaline earth metals.** They are also reactive elements with low densities, but generally less reactive and more dense than alkali metals.
- Groups 3–12 are the **outer transition metals** (or just "transition metals"). Their properties vary greatly, but they're generally hard, have high melting and boiling points, and fairly low reactivities.

- The two rows at the bottom of the periodic table are the **inner transition metals** and can't be said to be in any of the 18 groups of the periodic table. The top row is the lanthanides; these are shiny, reactive metals. The bottom row is the actinides; they are radioactive elements with a wide variety of uses.
- Group 17 is the **halogens.** These very highly reactive nonmetals combine readily with metals to form ionic compounds (see Chapter 3). The halogens are **diatomic elements,** which means they have the general formula X_2 (e.g., fluorine is F_2). Fluorine and chlorine are gases under standard conditions, bromine is a liquid, and iodine is a solid. Little is known of astatine because it is radioactive and short-lived.
- Group 18 is the **noble gases.** The noble gases are almost entirely unreactive because they have completely filled s- and p-orbitals, making them very stable. As we shall see shortly, all other elements tend to react so their orbitals fill, and they have the same number of electrons as the nearest noble gas.
- Hydrogen is the weirdo of the periodic table, having properties unlike any other element. Though it's found among other metals, it is a nonmetallic, diatomic gas. Hydrogen reacts slowly with other elements at room temperature, but may react blindingly fast when heated or catalyzed.

Periodic Trends

As previously mentioned, elements tend to react so they will end up with completely filled s- and p-orbitals to be like the noble gases. Because that's the most important idea in chemistry, let's really emphasize this point:

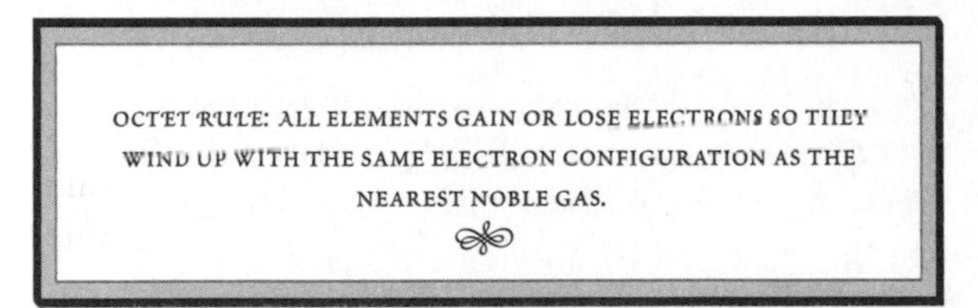

The octet rule is the driving force for chemical reactions and properties. If you learn only one thing from this chapter, learn the octet rule!

Elements tend to follow the **octet rule** because having filled s- and p-orbitals makes elements very stable. As a result, elements will tend to undergo chemical reactions in a way so that they end up with this electron configuration.

The Mole Says

The octet rule has that name because elements with the same electron configuration as the nearest noble gases usually have eight electrons in their highest energy level.

This explains why the alkali metals (group 1) are so reactive. Because they have only one more electron than the noble gases, they tend to react violently in ways that will get rid of this electron. Likewise, the halogens (group 17) have one fewer electron than the noble gases, which causes them to react violently to gain that electron. Generally, the closer the elements are to the noble gases on the periodic table, the more reactive they are.

Because this rule works so well, we can use it to predict how elements will behave based on their locations in the periodic table. Any property that can be predicted by an element's position on the periodic table is called a **periodic trend.**

Bad Reactions

Chemistry instructors generally get annoyed when students say that "nitrogen wants to …" in reference to a chemical reaction. After all, elements don't "want" to do anything. During the course of this book, when I say that an atom "wants to do" something, what I really mean is that the atom "will become more stable when it does" something.

Ionization Energy

The **ionization energy** of an element is the amount of energy required to pull one electron off an atom.

This ionization process is described by the following figure.

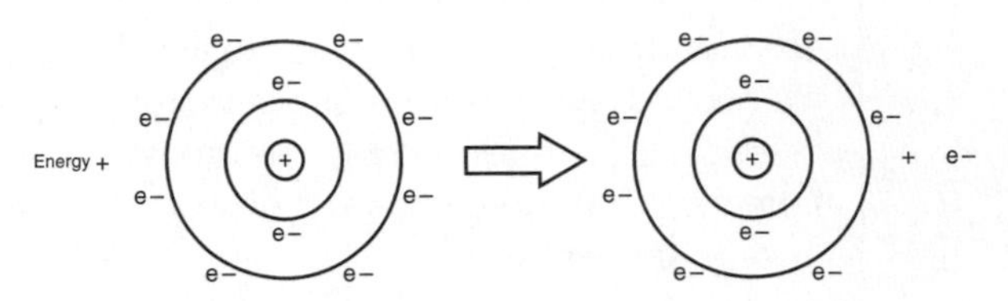

When enough energy is added to an atom, an electron can be removed in a process called ionization.

Elements on the left side want to lose electrons to be like the nearest noble gases, so they have very low ionization energies. Elements to the right, on the other hand, have high ionization energies because they want to gain electrons to be like the nearest noble gases. The highest ionization energies belong to the noble gases, because they're already stable with their current number of electrons.

Ionization energy decreases as you move down a family of elements in the periodic table. This is because electrons in low-energy levels repel electrons in higher-energy levels away from the nucleus in a phenomenon known as the *shielding effect.* Because of this, outer electrons are less tightly held to the nucleus than inner electrons, making them easier to pull off.

Molecular Meanings

The **shielding effect** is the tendency of electrons in inner energy levels to shield the nucleus from electrons in outer energy levels. This force repels outer electrons from the nucleus and makes it easier for them to be removed.

Electron Affinity

Electron affinity is the energy change that occurs when a gaseous atom picks up an extra electron. Elements of the left side of the periodic table want to lose electrons to be like the nearest noble gas, not gain them. As a result, not much energy is released when they pick up an extra electron, causing the electron affinity to be only slightly negative. However, elements on the right side of the periodic table very much want to gain extra electrons to be like the noble gases, which causes a great deal of energy to be released. Generally, the electron affinity of elements becomes more negative as you move from left to right across the periodic table. The exception to this is the noble gases—because they're already completely stable, they don't pick up electrons at all and have no electron affinity.

As you move down a family in the periodic table, elements want to gain electrons less because of the shielding effect. As a result, elements at the bottom of the periodic table tend to have less negative electron affinities than those at the top.

Electronegativity

Electronegativity is similar to electron affinity in that both measure how easy it is for an atom to gain electrons. However, while electron affinity deals with isolated atoms in the gas phase, electronegativity is a measure of how much an atom will pull electrons away from other atoms it has bonded to.

Because elements on the left side of the periodic table prefer to lose electrons and elements on the right side prefer to gain electrons, the trend for electronegativity increases as you move from left to right across the periodic table. Remember, however, that the noble gases have no electron affinity because they don't have any desire to gain electrons. Likewise, the shielding effect causes the electronegativity of elements to decrease as you move down a family in the periodic table.

Atomic Radius

The **atomic radius** is equal to one-half the distance between the nuclei of two bonded atoms of the same element. Though it's common to think of atoms as being spherical (or roughly so), quantum mechanics says they don't have any defined outer boundaries, so the definition we use defines this boundary, as shown in the following figure.

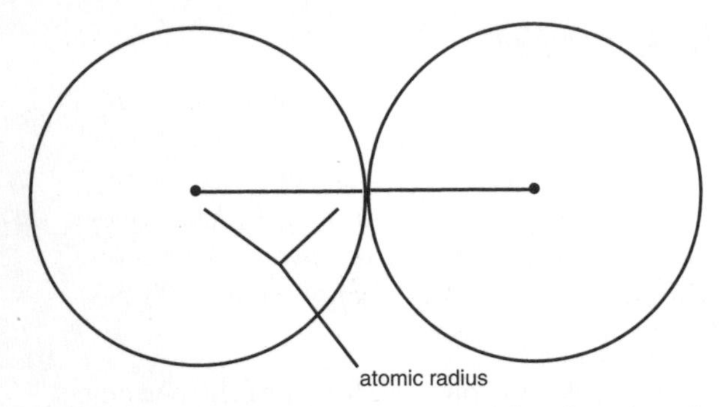

Because both atoms are the same, we can define the radius of each atom as half the distance between the two of them.

As you move from left to right across the periodic table, atomic radius decreases. This is because the outer electrons for each element have roughly the same energy, but the nucleus has greater positive charge, causing these electrons to be pulled in more tightly. As you move down a family on the periodic table, the atomic radius increases because every row has one more energy level than the one above it. Because the outer electrons have greater energies than the inner electrons, they're better able to pull away from the nucleus.

Ionic and Covalent Compounds

In This Chapter

- Characteristics and naming of ionic compounds
- Bonding in and properties of covalent compounds
- An introduction to hybrid orbitals
- Lewis structures and VSEPR theory

If you learned nothing else from Chapter 2, the octet rule should be embedded permanently in your mind: All elements tend to react so they have the same number of electrons as the nearest noble gas. This rule is going to come in handy in this chapter because it explains the two methods by which chemical compounds form. Let's take a look.

Ionic Compounds

Imagine an atom of lithium. It's got three electrons, one more than neon, the nearest noble gas.

As a result, it would love to give up that extra electron. Now, his buddy is an atom of chlorine. He's got 17 electrons, one fewer than argon, his nearest noble gas. As a result, he'll want to gain one more electron. This is a match made in heaven!

The Mole Says

From Chapter 2, remember that the tendency to want to gain electrons is known as electronegativity. Elements on the left side of the periodic table tend to want to lose electrons to be like the nearest noble gas, while elements on the right tend to want to gain electrons.

When these two atoms come near each other, the electronegative chlorine atom will want to take away lithium's extra electron. This causes chlorine to have more negative charge than positive charge (17 protons – 18 electrons = –1 charge) and the lithium to have more positive than negative charge (11 protons – 10 electrons = +1 charge). Because oppositely charged particles attract one another, they'll tend to stick to each other. The resulting compound is referred to as an *ionic compound*, because charged atoms or groups of atoms are referred to as *ions*.

Molecular Meanings

Ions are atoms or groups of atoms that have charge. If they have lost electrons so they have positive charge they are called **cations,** and if they have gained electrons so they have negative charge they're called **anions.** When an anion sticks to a cation, the result is an **ionic compound** (also called a **salt**).

Generally, ionic compounds are formed when two elements with vastly different electronegativities (greater than 2.1) bond with each other. Because metals and nonmetals frequently have such dissimilar electronegativities, it's usually a good guess that compounds formed by the combination of a metal and nonmetal are ionic.

Properties of Ionic Compounds

Because all ionic compounds are formed when anions and cations are attracted to one another, ionic compounds frequently have similar characteristics.

- **Ionic compounds form crystals.** Because ionic compounds consist of many positively and negatively charged ions stuck near one another, they tend to stack up in regular patterns that keep oppositely charged particles close to one another. These regular patterns of ions are referred to as **crystals,** as seen in the following figure.

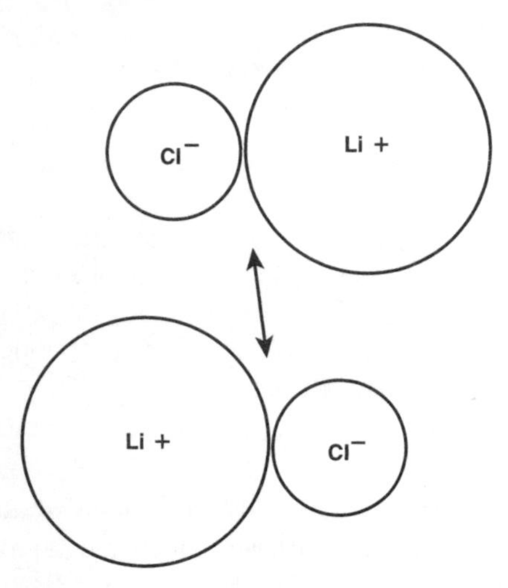

The positive charge on the lithium cation of one pair will be attracted to the negative charge of the chloride anion of the other pair.

- **Ionic compounds often have high melting and boiling temperatures.** To make an ionic compound melt, you need to put enough energy into the compound to make all of the cations and anions move away from one another. Because the interactions between all the ions are very strong, it takes a lot of energy to make this happen.
- **Ionic compounds are hard and brittle.** In ionic crystals, the cations and anions are tightly locked in place, which keeps them from moving around much; as a result, ionic compounds are hard. Likewise, ionic compounds are brittle because the addition of

force in the right places tends to force the entire crystal apart, rather than causing it to bend.

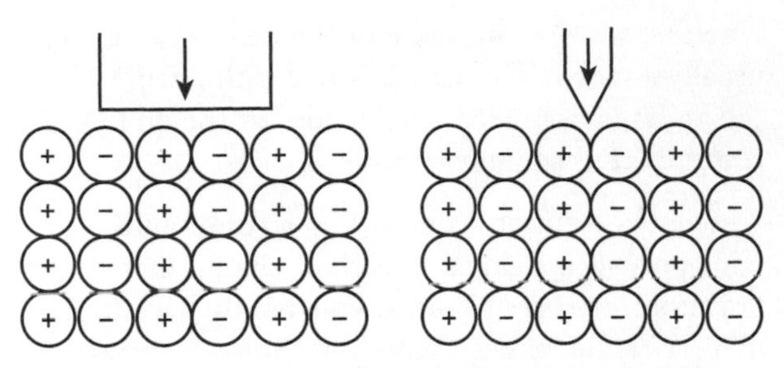

By applying force in a way that pries the cations and anions apart from one another, you can cause the crystal to break apart. The places where this can occur are called cleavage planes.

- **Ionic compounds conduct electricity when dissolved in water or melted.** Pure water is a very poor conductor of electricity. However, when salts dissolve in water they break up into their constituent cations and anions. It is the presence of these mobile ions that causes the water to conduct electricity. Likewise, when you melt an ionic compound, the ions are also able to move around freely, so the melted compound is also a good conductor.

Naming Ionic Compounds

One of the toughest things for chemistry students to learn is how to name ionic compounds.

Fortunately, there's an easy way to learn to do it.

Step 1: Determine the base name of the compound, which will always contain two words.

The first word is the name of the cation, which is usually a metal. The second word is the name of the anion (a nonmetal) with "-ide" at the end. As a result, NaF is sodium fluoride.

The only exceptions come when a compound contains **polyatomic ions,** ions that contain more than one atom. In cases like this, you simply have to memorize the names of the ions. Some common polyatomic ions include the acetate ion ($C_2H_3O_2^{-1}$), the ammonium ion (NH_4^{+1}), the bicarbonate ion (HCO_3^{-1}), the carbonate ion (CO_3^{-2}), the cyanide ion (CN^{-1}), the hydroxide ion (OH^{-1}), the nitrate ion (NO_3^{-1}), the phosphate ion (PO_4^{-3}), and the sulfate ion (SO_4^{-2}). As a result, the compound $NaNO_3$ is called sodium nitrate and $(NH_4)_2SO_4$ is called sodium sulfate.

Step 2: Determine whether or not the name of the compound requires a Roman numeral.

For many compounds, you can stop with the base name we described in Step 1. However, some elements form cations that have more than one possible charge. For example, both $FeCl_2$ and $FeCl_3$ are compounds that can be named iron chloride. To distinguish these compounds, we write a Roman numeral between the name of the cation and the anion to indicate the positive charge on the cation for atoms that can have more than one possible charge. Common elements that require Roman numerals are titanium, chromium, iron, nickel, copper, tin, and lead (though there are others, mostly transition metals).

To determine the Roman numeral required in these cases, use the following formula:

$$\text{Roman numeral} = \frac{-(\text{charge on anion}) \times (\text{number of anions})}{\text{number of cations}}$$

For example, let's determine thc formula of $FeCl_3$. The charge on the anion is –1, because chlorine wants to gain one electron to be like argon. The number of anions from the formula is 3, and the number of cations is 1. Plugging this into the equation, we find that the Roman numeral is $(-(-1 \times 3)/1) = 3$. This compound, then, is named iron(III) chloride. Likewise, FeO is iron(II) oxide (because $(-(-2 \times 1)/1) = 2$) and $Pb(SO_4)_2$ is lead(IV) oxide (because $(-(-2 \times 2)/1) = 4$).

Writing Ionic Formulas from Names

This is basically just the opposite of the process we discussed previously. Let's use the example calcium chloride:

- To start, let's figure out the formula and charge of both the cation and the anion. Because calcium needs to lose two electrons to be like its nearest noble gas, it has a charge of +2, giving us the formula Ca^{+2}. Because chlorine needs to gain one electron to be like argon, it has a charge of –1, giving us the formula Cl^{-1}.

- Devise a formula that gives the compound a neutral charge. Because the negative charge of chlorine is half that of calcium, you'll need two chlorine atoms to offset the positive charge of one calcium atom. The formula, then, is $CaCl_2$, indicating that two chlorine atoms have combined with one calcium atom.

The Mole Says

If you find you have more than one polyatomic ion when writing the formula of a chemical compound, you must always put parentheses around it before writing the number that tells you how many ions are present. Beryllium hydroxide is correctly written as $Be(OH)_2$, not $BeOH_2$. However, if the ion is not polyatomic, *never* use parentheses!

Covalent Compounds

When we discussed ionic compounds, we found that if the elements have very different electronegativities, the more electronegative atom will take electrons from the less electronegative one so they both have the same number of electrons as the nearest noble gases.

What this doesn't do is tell us what happens when two elements with similar electronegativities bond to each other. If both atoms want to grab electrons, they can both only be happy if they share.

Before we go too much farther with this explanation, let's go over a couple of key terms:

- **Valence electrons** are the number of s- and p-electrons that an element has past that last noble gas. For example, lithium has one valence electron and nitrogen has five valence electrons. When elements bond by sharing electrons, they generally want to have a total of eight valence electrons; the main exception to this rule is hydrogen, which only wants two.
- **Covalent bonds** are bonds formed when two atoms share a pair of valence electrons. All covalent bonds contain two electrons. Because nonmetals usually have high electronegativities, they form covalent bonds with one another.

Let's see how this works with a common example, water:

Oxygen has six valence electrons, making it two short of the eight it needs to be like neon, its closest noble gas. Every hydrogen atom has one valence electron, making it one short of the two it needs to be like helium. Here's the cool part: If hydrogen and oxygen share their valence electrons, they can both pretend that they have the same number of valence electrons as their nearest noble gases, as shown in the following figure.

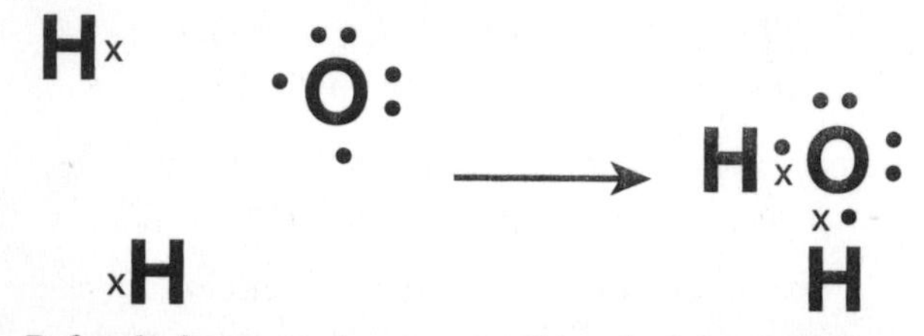

Before hydrogen and oxygen combine, both are missing valence electrons. However, after they share their unpaired electrons with one another, both have the right number of valence electrons. Each shared pair of valence electrons is called a covalent bond.

The Mole Says

You may have noticed that this figure initially shows the electrons spread out on all four sides of oxygen rather than paired in three positions. Whenever possible, we show electrons unpaired because electrons repel one another and only pair up when there is no alternative.

As you can see, both atoms now have their desired number of valence electrons. Each hydrogen has two valence electrons next to it (including an electron from oxygen it is sharing) and oxygen has eight valence electrons next to it (including two electrons, one shared with each hydrogen). Because both elements have the same number of valence electrons as their nearest noble gas, they form a stable covalent compound, H_2O.

The Mole Says

Many atoms in covalent molecules have pairs of electrons that are not shared with other atoms. These are usually referred to as lone pairs or nonbonding electron pairs.

In some cases, two nonmetals can covalently bond more than once. Let's use O_2 as an example. Both oxygen atoms have six valence electrons, and both atoms need two more valence electrons to be like neon. Fortunately, by combining both sets of unpaired electrons simultaneously, they both achieve their desired electron configurations:

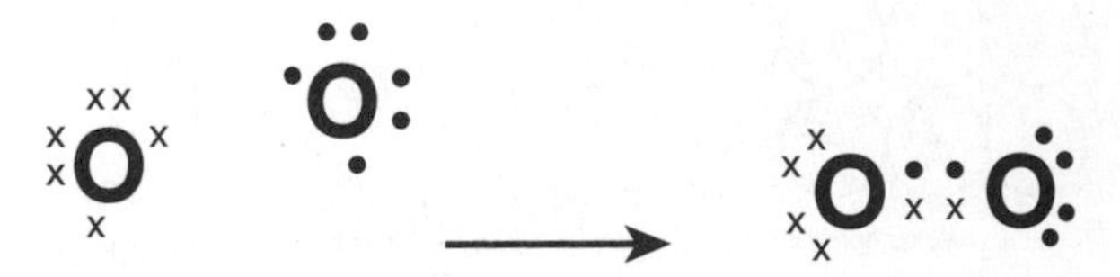

By combining more than one unpaired electron at a time, a double bond is formed and both oxygen atoms end up with eight valence electrons.

Two atoms can bond to each other a maximum of three times (as in N_2). However, don't assume that if an atom forms a multiple bond in one molecule that it always forms a multiple bond. Every molecule is unique and requires separate analysis.

Properties of Covalent Compounds

Like ionic compounds, covalent compounds share many properties because of the similar way that they bond. However, before we discuss these properties, we should note that covalent compounds have little in common with ionic compounds, except that both are the result of atoms following the octet rule.

To illustrate this concept, take a look at the following figure:

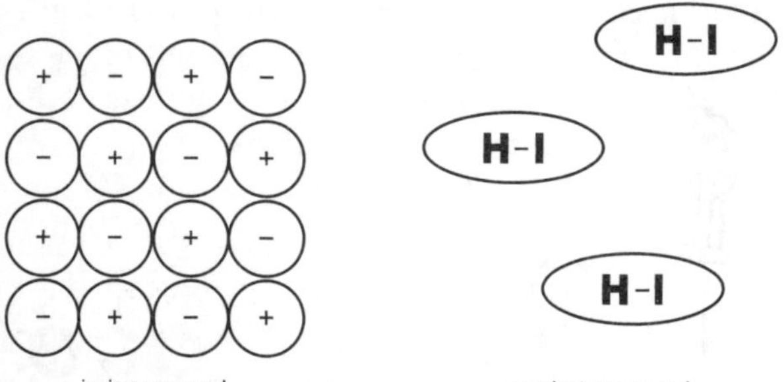

The properties of solid ionic compounds are based on the fact that the ions are rigidly held in place by electrostatic forces. Molecules in covalent compounds, however, operate with relative independence from neighboring molecules.

Unlike ionic compounds, where all of the ions in a large crystal are stuck together in a big chunk because of electrostatic forces, the molecules in a covalent compound are only stuck to each other by much weaker intermolecular forces (more about this in Chapter 5). As a result, the molecules in a covalent compound are not attracted to each other nearly as strongly as the ions in an ionic compound.

This difference in structure is vital to understanding the differences between the properties of ionic and covalent compounds:

- Covalent compounds usually have low melting and boiling temperatures relative to ionic compounds. Unlike ionic compounds, which have high melting and boiling temperatures because all of the cations and anions are stuck tightly together, covalent compounds generally have low melting and boiling points because it doesn't take much energy to pull molecules apart from each other.

Bad Reactions

Many chemistry students believe that covalent bonds are broken when covalent compounds melt. *This is not so!* When a covalent compound melts, the molecules remain intact and are simply moved away from one another.

- Covalent compounds are poor conductors of electricity. Unlike ionic compounds, which have mobile ions when melted or dissolved in water, covalent compounds have no ions to help water conduct electricity. As a result, covalent compounds are excellent electrical insulators.
- Covalent compounds sometimes burn. By far the largest group of covalent compounds is

organic compounds, compounds that contain both carbon and hydrogen. When organic compounds are combined with oxygen at high temperatures, they generally burn very well. Although covalent compounds that don't contain carbon and hydrogen (called inorganic compounds) frequently don't burn, they're far outnumbered by organic compounds.

What's in a Name? Covalent Nomenclature

As with ionic compounds, covalent compounds have two-word names. The first name is the name of the first atom in the compound, and the second is the name of the second atom in the compound with "-ide" at the end. For example, HF is hydrogen fluoride.

If more than one atom of the element is present in the compound, you need to indicate the number of atoms with a prefix before the name of the element in the two-word name. The most common prefixes are shown in the following table.

Number of Atoms	Prefix
1	mono- (use only for oxygen)
2	di-
3	tri-
4	tetra-
5	penta-
6	hexa-

Using this convention, CF_4 is carbon tetrafluoride. It's as simple as that!

To come up with the formula of a compound from the name, simply reverse this process. For example, "carbon dioxide" refers to a molecule having one carbon atom and two oxygen atoms, CO_2.

The only exceptions to these naming conventions come with some very common molecules and with pure elements. Many molecules have common names you just need to memorize, among them water (H_2O), ammonia (NH_3), and methane (CH_4). Pure elements typically have the same formula as the name (aluminum is stated as Al) except for the diatomic elements oxygen (O_2), nitrogen (N_2), fluorine (F_2), chlorine (Cl_2), bromine (Br_2), iodine (I_2), and hydrogen (H_2).

The Mole Says

If you shade the diatomic elements on the periodic table, you'll find that six of them form the shape of a seven on the right side of the periodic table and the lone weirdo is hydrogen. As a result, it's easy to remember the seven diatomic elements as "the big seven and the weirdo."

Hybrid Orbitals

Way back in Chapter 1 we talked about orbitals and orbital filling diagrams. To refresh your memory, let's take a look at the orbital filling diagram for carbon:

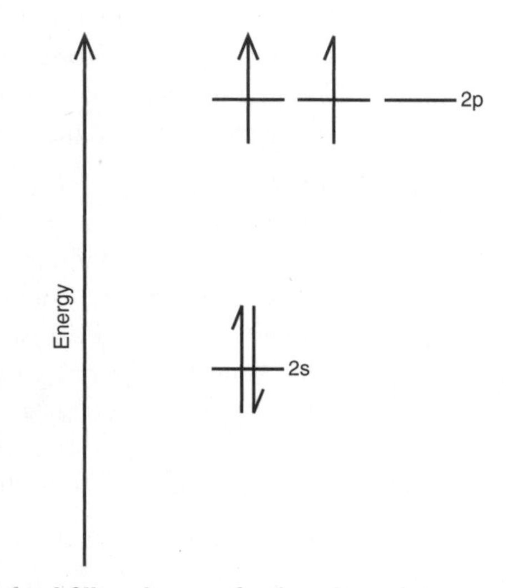

The orbital filling diagram for the valence electrons on carbon.

When we talked about covalent bonding earlier in this chapter, we mentioned that unpaired electrons on one atom pair up with unpaired electrons on the other atom to form a covalent bond. From the diagram above, however, we can't get the four covalent bonds that are necessary to form methane, CH_4.

Fortunately, carbon doesn't bond quite like this. What really happens when carbon bonds covalently with other elements is that these four orbitals mix with one another to form four identical hybridized orbitals. The names of these new hybridized orbitals are a combination of the names of the original four orbitals that made them. In our example, one s-orbital combines with three p-orbitals to form four sp^3 orbitals.

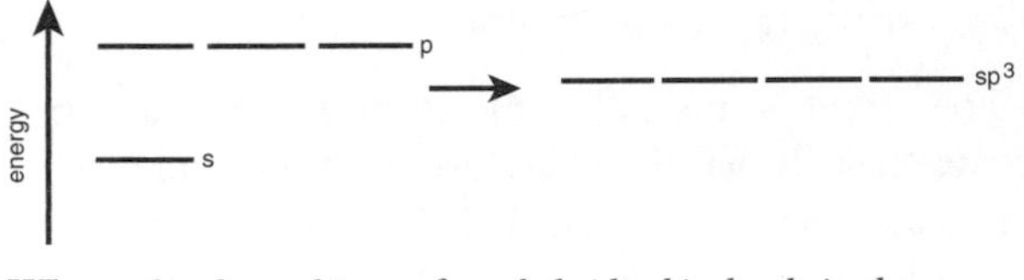

When orbitals combine to form hybrid orbitals, their shapes and energies are averaged.

The number of hybrid orbitals that form in a covalent molecule depend on the number of single bonds that an atom forms with other elements and the number of lone pairs of electrons present—both covalent bonds and lone pairs reside in hybrid orbitals. Multiple bonds occur when unhybridized p-orbitals overlap with one another.

The Mole Says

Lone pairs of electrons and electrons involved in single covalent bonds reside in hybridized orbitals. Multiple bonds reside in unused p-orbitals from the two atoms that overlap. Single bonds are referred to as **sigma bonds** and multiple bonds are referred to as **pi bonds**.

Lewis Structures

Lewis structures are nothing more than pictures of covalent molecules that show where all of the atoms and valence electrons are. Fortunately, these are usually fairly easy to draw.

The secret is this: All of the atoms will bond in a pattern such that the unbonded electrons will match up with one another. Aside from that, you can do pretty much whatever you'd like to connect all of the atoms in a molecule, as long as you follow the rules below:

- All atoms need to bond the "correct" number of times. Generally, in neutral molecules, halogens and hydrogen bond once, oxygen's family bonds twice, nitrogen's family and boron bond three times, and carbon's family bonds four times. In charged molecules, the only difference is that oxygen's family may bond once or twice and nitrogen's family may bond two, three, or four times.
- All atoms will need to have the same number of valence electrons as the nearest noble gas when they're done bonding. Most elements need eight electrons, while hydrogen needs two. Boron is the weirdo in this case, needing six electrons in neutral compounds and eight electrons when it has a negative charge.
- All valence electrons need to be accounted for. If you're trying to draw the Lewis structure of carbon dioxide, your final drawing must contain 16 electrons (6 from each

oxygen atom and 4 from carbon). Any electrons that aren't involved in bonding must be added as lone pairs around atoms that need them.

- For charged molecules, you need to determine the charge on each atom by using the following formula: The number of lone pair electrons plus the number of bonds, subtracted by the number of valence electrons the atom normally has, determines the charge on the atom.

And that's all you need to do. As an example, if you're trying to find the Lewis structure for carbon dioxide, you'll find the only possible structure that gives carbon four bonds and each oxygen two bonds is:

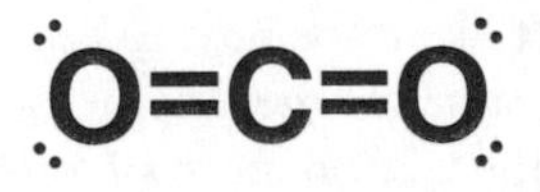

This is the only Lewis structure that gives all of the atoms the right number of bonds and electrons.

As you can see, unbonded electron pairs have been added to oxygen so that each oxygen atom has the eight valence electrons it wants. Remember, each bond counts as two electrons!

The Mole Says

Sometimes it's possible to draw more than one arrangement of bonds and atoms for a particular formula. If the atoms are in the same place but the bonds and electrons have moved, the two molecules are referred to as **resonance structures.** If the atoms are bonded in fundamentally different patterns, the two molecules are **isomers.**

Introducing VSEPR Theory

VSEPR stands for "Valence Shell Electron Pair Repulsion," which sounds pretty awful. Fortunately, VSEPR (pronounced *vesper*) theory is easy to understand.

Electrons don't like to be next to each other because they have the same charge. If you look at the electrons around the central atom of a molecule, you'll find that they arrange themselves so they're as far away from each other as possible.

As a result, you can tell the shape, bond angles, and bond hybridizations of a molecule from its Lewis structure. A flow chart allowing us to do exactly that is shown next.

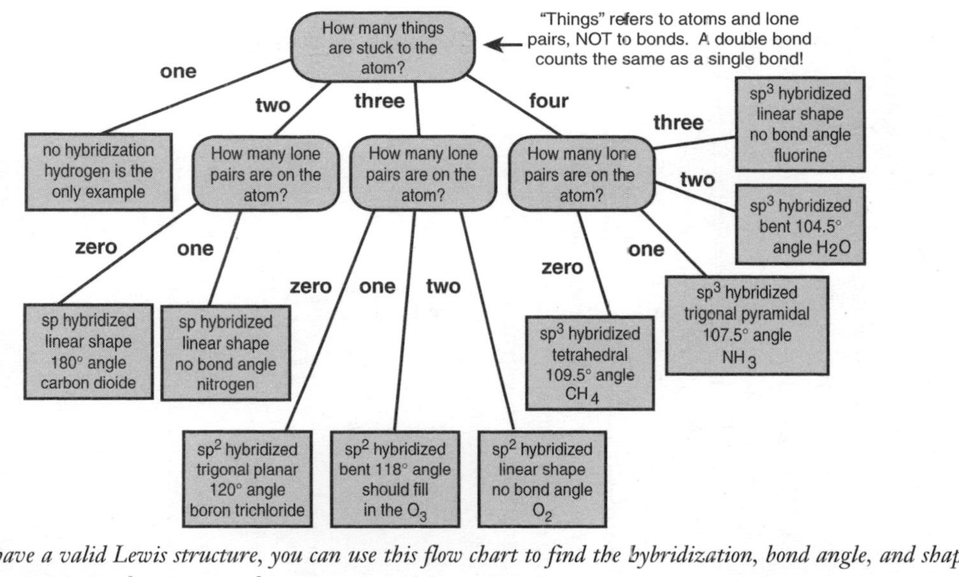

Once you have a valid Lewis structure, you can use this flow chart to find the hybridization, bond angle, and shape of any atom on any covalent compound.

How to use this chart:

- At the top, you're asked, "How many things are stuck to the atom?" In this question, *things* refers to atoms and lone pairs, but not to the bonds themselves.
- The second question asks, "How many lone pairs are on the atom?" This question simply asks us to count how many of the *things* from the first question are atoms.

Molecules that contain lone pairs of electrons tend to have smaller bond angles than atoms without lone pairs because the relatively large lone pair electrons tend to squish the other atoms together. This is shown in the next figure.

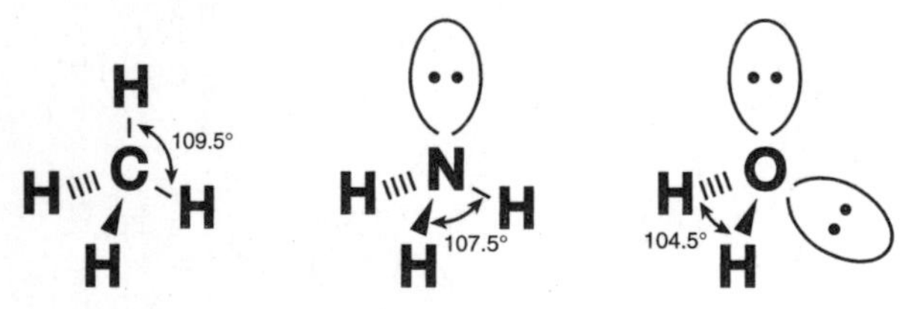

Though all three of these molecules exhibit sp^3 hybridization around the central atom, the bond angles get successively smaller because of the increasing numbers of lone pairs.

Moles, Equations, and Stoichiometry

In This Chapter

- Using the mole in chemistry
- The care and feeding of chemical equations
- Stoichiometry calculations

The most common thing that chemistry instructors hear from students is, "I liked this class until we had to learn all that math!" Fortunately, you've got nothing to worry about. Although this chapter involves a lot of plugging numbers into your calculator, the calculations are easy if you know the secret chemical tricks.

The Mole

It's not at all scary if I tell you that I'm wearing a "pair" of shoes because all it means is that I'm wearing two shoes. If I tell you that I bought a "dozen" eggs at the market, you know that I bought 12 eggs.

However, if I say I have a *mole* of copper, you may get nervous. Never fear—just as the words "pair" and "dozen" are shorthand for numbers used in everyday life, the word "mole" is just shorthand for a number that comes up quite a bit in chemistry. This number is 6.02×10^{23}, and is frequently referred to as Avogadro's number. As a result, if I say I have a mole of copper, this means I have 6.02×10^{23} atoms of copper. Similarly, if I were to say that I had a mole of guitars, that would mean that I had 6.02×10^{23} guitars.

Molecular Meanings

A **mole** is equal to 6.02×10^{23} things. Though you could, in theory, have a mole of anything, this number is so huge that we usually only speak of having moles of atoms or molecules.

Molar Mass

The mole is useful because it's the main unit used in chemical equations. Unfortunately, it's not very convenient to count out 6.02×10^{23} molecules, so it's nice to be able to convert moles to something more easily measured, like grams. The conversion factor for this calculation is called the *molar mass*, which is equal to the weight of one mole of an element or chemical compound. The units for molar mass are expressed as grams/mole (g/mol).

Molecular Meanings

The **molar mass** of a substance is the weight of 6.02×10^{23} atoms or molecules of that material in grams. The unit of molar mass is given as grams/mole, abbreviated as g/mol. Other synonymous terms for molar mass are **molecular mass, molecular weight,** and **gram formula weight.**

To find the molar mass of a chemical compound, add together all of the atomic masses of the elements in the compound, using the periodic table. For example, the molar mass of water, H_2O, is found by adding the weight of two atoms of hydrogen (2×1.01 amu) to the weight of one atom of oxygen (1×16.00 amu), giving us 18.02 amu. Because atomic mass units are to atoms what g/mol are to molar mass, the molar mass of water is 18.02 g/mol.

Converting Between Moles, Molecules, and Grams

Once you know how to find the molar mass of a substance, you can convert between moles, grams, and molecules of that substance. To do so, use the following figure.

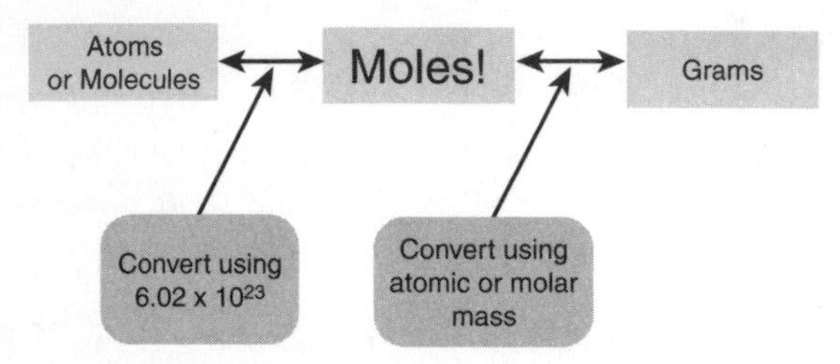

This map can be used to convert between grams, molecules, and moles of any chemical compound. When converting from moles to either grams or molecules, multiply by the conversion factor. When converting from grams or molecules to moles, divide by the conversion factor.

The diagram above gives you all of the information you need to convert between moles, grams, and molecules. Basically, all you have to do is figure out where you're starting on this chart and where you're trying to end up (for example, if you want to figure out how many molecules are in 75.4 grams of water, you'd start at the "grams" box and end up at the "Atoms or molecules" box). The conversion factors you need for each step are given on the line between the boxes you're moving between. These conversion factors are used in the following ways:

- If you're converting from moles to something else, you'll multiply the number of moles by the conversion factor.
- If you're converting to moles from something else, you'll divide the number you're converting by the given conversion factor.

Let's convert 75.0 grams of water to molecules.

From the preceding diagram, we can see that to go from grams to molecules, we must first convert grams to moles. To do this we need the molar mass as our conversion factor—in this case, 18.01 g/mol. Because we're converting from grams to moles, we divide the number of grams by the molar mass, giving us (75.0 grams ÷ 18.01 g/mol) = 4.16 moles.

To go from moles to molecules, the map shows us that we need to multiply this value by the conversion factor 6.02×10^{23}. Performing this calculation, we find that we have (4.16 moles $\times 6.02 \times 10^{23}$ molecules) = 2.51×10^{24} molecules of water. It's as simple as that!

Chemical Equations

Chemical equations are recipes for telling us how to do chemical reactions. Unlike recipes written in plain English (or whatever language you cook in), chemical equations are written in condensed shorthand that chemists can easily understand. In this section, we'll learn how to decode these chemical recipes and how to write some of our own.

The Basic Features of an Equation

Chemical equations, while seemingly simple, contain a great deal of information to the chemically informed mind. Let's take a look at some of the most common features you may run into while reading and writing equations, using the equation for the formation of water as an example:

$$2\ H_{2(g)} + O_{2(g)} \xrightarrow{\Delta} 2\ H_2O_{(g)}$$

- The basic form of the equation shows that the elements or compounds to the left of the arrow (the reactants or reagents) will form, under certain conditions, the elements or compounds to the right of the arrow (the products).
- The numbers in front of each chemical formula tell us how many moles of each compound will either be needed or formed in the reaction. We can determine these numbers by a process known as **balancing equations,** which we'll learn shortly.
- The small terms written as subscripts under each of the products and reactants indicate the physical state of the products and reactants in a reaction. The subscript (s) indicates a solid, (l) indicates a liquid, (g) indicates a gas, and (aq) indicates that the compound is dissolved in water.
- There are sometimes symbols written above or below the arrow in a chemical reaction to indicate the procedures that need to be followed to make the reaction occur. Common terms include the following: Δ indicates that energy needs to be added to the reactants for the reaction to occur; any temperature

or pressure indicates the reaction conditions; time values indicate the length of time the reaction needs to proceed; chemical formulas indicate that a chemical will need to be added to the reactants at some point, or that the chemical should be used as the solvent.

The Mole Says

Though all of these *may* be present in a chemical equation, it may not be necessary to specify the specific states of the products and reactants or to indicate the reaction conditions.

Balancing Equations

Chemical equations have numbers called **coefficients** in front of each chemical formula indicating the number of molecules of each compound or element needed. This follows from the law of conservation of mass (Chapter 1), which states that the amount of stuff you make in any process will be the same as the amount of stuff you started with. Because the "stuff" in this case is atoms of reactants and products, we need to add coefficients to the equations that will make the equations follow the law of conservation of mass.

The Mole Says

The law of conservation of mass requires that the numbers of atoms on either side of the equation be the same, but *not* that the number of molecules on either side of the equation be the same. As a result, don't worry if the sum of the coefficients on one side of the equation isn't the same as on the other side—as long as you have the same number of atoms, you're in good shape!

Let's balance the unbalanced equation $N_2 + H_2 \rightarrow NH_3$. To do this, we'll follow these steps:

- Make an inventory of the number of atoms on both sides of the equation. At this point, our reactants consist of two atoms of nitrogen and two atoms of hydrogen; our products consist of one atom of nitrogen and three atoms of hydrogen.
- Change a coefficient to make the number of atoms of one element the same on both sides of the equation. In our example, there are two atoms of nitrogen on the reactant side and one atom of nitrogen on the product side. To make the number of nitrogen atoms in the products equal to the number of nitrogen atoms in the reactants, write a "2" in front of NH_3 in the equation.

- Keep redoing these steps until the number of atoms of each element is the same on both sides of the equation. Don't worry if this takes a bunch of steps—you'll eventually get the final answer: $N_2 + 3\ H_2 \rightarrow 2\ NH_3$.

Because some students have trouble balancing equations at first, you may want to take note of the following hints/warnings:

- Never change the chemical formulas in the equation. If you do, you are guaranteed to get the wrong answer!
- If you've changed the coefficients many times and still haven't gotten an answer, start the problem over again. If that doesn't work, start again by putting a "2" in front of the most complicated-looking molecule. Oddly enough, this can help solve the most annoying problems.
- If all the coefficients in your final equation can be divided by a lowest common denominator (as in the case $2\ N_2 + 6\ H_2 \rightarrow 4\ NH_3$), divide them all to get the lowest possible coefficients.

Types of Chemical Reactions

Just as there are many types of recipes, there are many types of chemical reactions. These include the following:

- **Combustion reactions** occur when organic molecules (molecules containing carbon and hydrogen) combine with oxygen to form carbon dioxide, water, and heat.
- **Synthesis reactions** occur when small molecules combine to form more complex ones.
- **Decomposition reactions** occur when complex molecules break apart to form simpler ones.
- **Single displacement reactions** occur when a pure element switches places with one of the elements in a chemical compound. These are the simplest type of redox reactions, as we'll discuss in Chapter 11.
- **Double displacement reactions** occur when the cations of two ionic compounds switch places.
- **Acid-base reactions** are essentially double displacement reactions that result in the formation of water. These typically occur when acids (compounds that give off H^+ ions) combine with bases (compounds that contain OH^- ions). More about these in Chapter 10.

Knowing your types of reaction can help you to predict the products of a chemical reaction, if that information isn't given to you. For example, if you've got a reaction in which two pure elements combine with one another, it's probably safe to predict that the result will be a synthesis reaction.

Stoichiometry

Now that we've learned about moles and equations, it's time to put it all together and learn what it's good for. The subject of **stoichiometry** (pronounced *stoy-key-AH-meh-tree*) is a way of figuring out how much of a product can be made from a given quantity of reactant.

The idea is simple. Let's say that the perfect formula for ice water is five ice cubes for every eight ounces of water. Given this information, how many perfect glasses of ice water can I make given 10 ice cubes and an excess quantity of water? If you guessed two (and I imagine you did), you're already a stoichiometry whiz!

The Mole Says

It's frequent in stoichiometry calculations to use the term **excess quantity** when referring to one of the reactants. This simply means that we have a larger-than-needed amount of the excess reactant and a smaller quantity of the other one (called the **limiting reactant,** as we'll see soon). As a result, the amount of product that can be formed depends on the other reactant, not the excess one.

Now that you've done your first stoichiometric calculation with ice water, it's time to move on to the sorts of questions that chemists like to ask. Here's one now:

Example: Using the equation $2\ H_2 + O_2 \rightarrow 2\ H_2O$, determine how many grams of water can be formed from 10.0 grams of oxygen and an excess of hydrogen.

You've probably already figured out that this isn't much like the ice water example. Fortunately, we've got a handy diagram to help us figure out what to do:

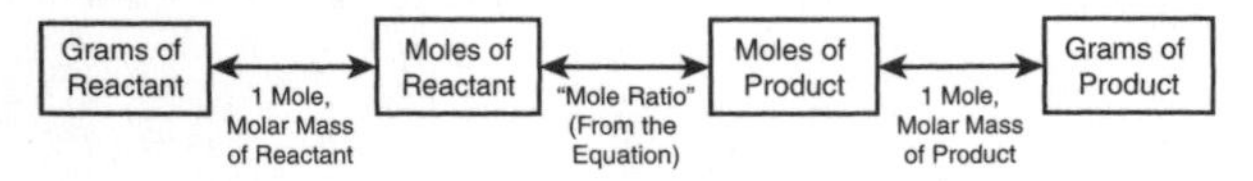

This handy chart will help you with mass-to-mass or mole-to-mole stoichiometry problems, like the one in our example.

To use this diagram, follow these steps:

1. Find the box that corresponds to the information you were given in the problem. In our case, we were given "10.0 grams of oxygen," so we'll start at the "Grams of Reactant" box.
2. To solve this or any other stoichiometry problem, you'll do a series of calculations until you move across the diagram to the desired destination box. In our example, that's the "Grams of Product" box, because we're trying to find the number of grams of water that can be formed.

3. Our first calculation takes us from the "Grams of Reactant" box to the "Moles of Reactant" box. Because we learned earlier in this chapter how to convert between grams and moles using the molar mass, we know that we need only divide 10.0 grams of oxygen by oxygen's molar mass (32.00 g/mol) to get the answer: 0.313 moles of oxygen.
4. Our second calculation takes us from the "Moles of Reactant" box to the "Moles of Product" box, multiplying by the mole ratio. For stoichiometric calculations, the mole ratio is equal to the coefficient in front of the product over the coefficient in front of the reactant. In our example, this is equal to the number in front of water, 2, over the number in front of oxygen, 1. Since 2/1 = 2, the mole ratio is 2. When we multiply the mole ratio by the number of moles of oxygen we had from step 3, we end up with 0.626 moles of water.
5. Our third and final calculation takes us from "Moles of Product" to "Grams of Product." Again, we know from earlier in this chapter that we need only multiply the number of moles by the molar mass of water (18.02 g/mol) to find the number of grams of water formed: 11.3 grams.

It's as simple as that!

Limiting Reactants

Now that you're a pro at stoichiometry, we'll look at a much tougher example: Using the recipe for ice water we used before (five ice cubes for every eight ounces of water), determine how many glasses of ice water we can make from 10 ice cubes and 800 ounces of water.

If you came up with two as your answer, you're correct! Let's see how we do this conceptually:

- Using this recipe, we can make 2 glasses of water from 10 ice cubes.
- With the same recipe we can make 100 glasses of water from 800 ounces of water.
- Because we run out of ice before we run out of water, we can only make two glasses of water.

In this example, ice is our limiting reactant. The ice is said to be "limiting" because it is the ingredient we run out of first, which puts a limit on how much ice water we can make. The water is called the excess reactant because we have more of it than is needed.

For chemical examples, we do an identical calculation. Let's say that we're asked to figure out, using the equation $2\ H_2 + O_2 \rightarrow 2\ H_2O$, how many grams of water can be made from 15.0 grams of hydrogen and 45.0 grams of oxygen. If you do the stoichiometry calculations using the diagram we discussed earlier, you find that you can make 134 grams of water from 15.0 grams of hydrogen and 50.7 grams of water from 45.0 grams of oxygen.

The Mole Says

Whenever you do a limiting reactant problem, you need to do two stoichiometry calculations, one for each reactant. The answer that is the smaller of the two is the correct answer for the problem, and the reactant that resulted in the smaller number is the limiting reactant.

What this means is that we run out of oxygen before we run out of hydrogen. This makes oxygen the limiting reactant, and the amount of water that can be made is 50.7 grams.

Liquids, Solutions, and Solids

In This Chapter

- Intermolecular forces
- Concentration in solutions
- The types of solids

You've probably already noticed that atoms are very small. As a result, it's hard to know what they're thinking. Are they plotting to overthrow the government? Are they responsible for the return of swing dancing? Fortunately, chemistry provides some answers, and in this chapter we'll learn what molecules are thinking, whether they're solids, liquids, or dissolved in other compounds.

Liquids

A **liquid** is the form of matter in which molecules move around freely but still experience intermolecular forces. This definition does a pretty good job of explaining what you probably already know about liquids (e.g., they change shape but not volume, they pool up if you spill them, etc.).

The big question here is, "What's an intermolecular force?" Well, I'm glad you asked. **Intermolecular forces** are attractive forces between two different molecules. Intermolecular forces provide the attraction between covalent molecules that causes them to form a liquid, rather than fly apart as gases do.

Let's take a look at the three types of intermolecular force.

Dipole-Dipole Forces

When we talked about covalent compounds in Chapter 3, we mentioned that the atoms that form the molecules must have similar electronegativities so they can share electrons. However, what happens if the electronegativities are similar, but not exactly the same?

If you guessed that the electrons will spend more time around the more electronegative atom, give yourself a cookie, because that's exactly what happens. Let's take a look at an O-F bond, in which fluorine is more electronegative than oxygen:

$$\overset{\delta+}{\mathbf{O}} - \overset{\delta-}{\mathbf{F}}$$

$$\longmapsto$$

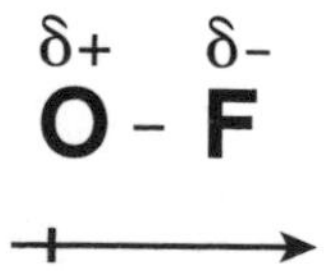

Because fluorine is more electronegative than oxygen, it pulls some of oxygen's electrons away from it, giving fluorine a partial negative charge.

Because fluorine is more electronegative than oxygen, the electrons in the O-F bond will tend to move toward the fluorine atom, giving it a partial negative charge, denoted by δ–. Likewise, since the electrons are being pulled away from oxygen, it has a partial positive charge, denoted by δ+. Covalent bonds where the electrons are distributed unevenly are referred to as *polar covalent bonds.* As in the diagram, dipole arrows point toward the electron with the partial negative charge.

Looking at the OF_2 molecule, we can see the effect this has on the overall distribution of electrons in the molecule:

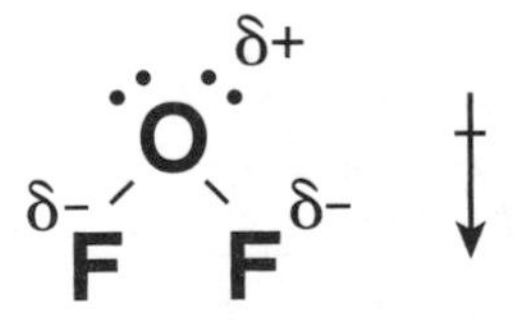

Both fluorine atoms pull electrons from oxygen, causing both of the O-F bonds to be polar and the whole OF_2 molecule to be polar.

Just as in the example of a single O-F bond, we see here that the electrons are pulled toward the side of the molecule containing the fluorine atoms. Because the electrons are unevenly distributed in the molecule, the molecule is referred to as a *polar molecule*.

In liquids containing polar molecules, the side of the molecule with partial positive charge will align itself with the partially negative sides of neighboring molecules. This attractive force is called a **dipole-dipole force.** While dipole-dipole forces are strong enough to keep the molecules in a liquid near each other, they're much weaker than ionic or covalent bonds. As a result, covalent molecules are able to move freely throughout the liquid.

Molecular Meanings

Polar covalent bonds are bonds in which electrons are preferentially pulled toward the more electronegative atom; **polar molecules** are covalent molecules in which electrons are preferentially pulled toward one region in the molecule. Not all molecules with polar covalent bonds are polar–if the polar bonds are distributed evenly throughout the molecule, the charges cancel each other out.

Hydrogen Bonding

Some very polar covalent compounds contain a hydrogen atom bonded to a nitrogen, fluorine, or oxygen atom. As a result, the hydrogen atom on one molecule (which has a high partial positive charge) has a very strong attraction to the lone pair of electrons on the N, F, or O atom on a neighboring molecule. This attraction is particularly strong because hydrogen has no inner electrons to shield the nucleus from other atoms. This very strong force is called a **hydrogen bond.** Hydrogen bonds, while still not as strong as covalent bonds or the attraction between cations and anions, are still much stronger than other intermolecular forces.

Bad Reactions

Hydrogen bonds are not real chemical bonds! Though hydrogen bonding is a particularly strong intermolecular force, it's still nowhere near as strong as a covalent bond.

London Dispersion Forces

So far, all of the intermolecular forces we've seen are between polar molecules in a liquid. You might be surprised to find that nonpolar molecules also use the attraction between opposite charges to stay together in a liquid. Since it's not obvious how molecules without any charges would use oppositely charged atoms to stick together, we'll need to investigate further.

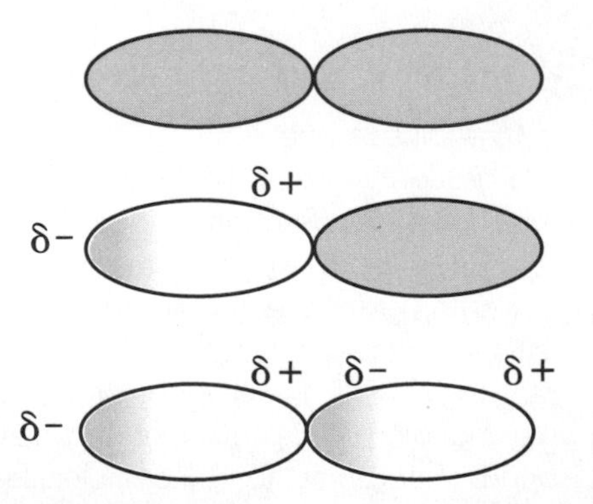

London dispersion forces are created when one molecule with a temporary dipole causes another to become temporarily polar.

In the top illustration in the preceding figure, we see two covalent molecules next to each other. As expected of nonpolar molecules, neither has a charge. However, if the electrons on one side of the molecule were to temporarily move to the other side, the molecule would become very temporarily polar, as shown in the second illustration. Because the molecule on the left is polar, the electrons in the molecule on the right will tend to move toward it, causing it to become polar. The attraction between these two temporarily polar molecules is called a **London dispersion force.**

As you might expect, this effect is temporary, because the random movements of electrons within a molecule quickly cause the temporary dipole to disappear. As a result, this very weak, very short-lived force is nowhere near as strong an interaction as hydrogen bonds or dipole-dipole forces.

The Mole Says

We talked about how hydrogen bonds are stronger than dipole-dipole forces, which are stronger than London dispersion forces. As it turns out, the strength of these forces in a molecule has a strong effect on the properties of the molecule. For example, water, which undergoes strong hydrogen bonding, has a much higher melting and boiling point than other molecules of similar weight.

Solutions

A **solution** is just another fancy word for "homogeneous mixture" (see Chapter 2). In solutions, one material (the **solute**) is completely dissolved in another (the **solvent**). Examples of solutions that you're probably familiar with include fruit punch and contact lens solution.

The Mole Says

Though we'll discuss solid solutes in liquid solvents, it's possible to have solutions with different phases. For example, if a liquid dissolves in a liquid, they're referred to as being **miscible,** and the major component is defined as the solvent. Likewise, air is a solution, though all of the components are gases.

One frequently asked question is, "How can I tell if a particular solute will dissolve in a solvent?" After all, it's clear that not all solids dissolve in liquids, because if they did, you'd burn holes in yourself every time you drank a glass of water.

The best way to tell if something will dissolve is to look at the polarities of the solvent and solute. Polar solutes tend to dissolve well in polar solvents and nonpolar solutes tend to dissolve well in nonpolar solvents ("like dissolves like"). The reason for this is that when solvent and solute have similar polarities, they tend to attract one another more than they attract other molecules of the same material.

Concentration

When talking about a solution, it's important to know how much of the solute is dissolved in the solvent. Let's explore some of the ways you can describe the **concentration** of a solution.

The amount of solute present in a solution can be described without numbers by one of the following terms:

- **Unsaturated solutions** are still capable of dissolving more solute. For example, if you dissolve one teaspoon of sugar in a glass of water, the solution is unsaturated because more sugar could still be dissolved.
- **Saturated solutions** have dissolved the maximum possible amount of solute. If you've ever seen a small child make Kool Aid, you know that the sludge on the bottom means that the maximum amount of powder has dissolved, and the solution is saturated.
- **Supersaturated solutions** have dissolved more than the usual maximum possible quantity of solute. For this reason, these solutions are unstable and the addition of a small mote of dust can cause crystals to form until the solution reaches saturation.

In addition to the qualitative methods of describing solubility, there are many, many ways of describing precisely how much solute is present in a solution. The following are the most commonly used terms, though by no means the only ones you might run into:

- **Molarity (M)** is the most common method for describing the concentration of a solution, and is defined as the number of moles of solute present per liter of solution. For example, if you wanted to determine the molarity of a solution in which 2.0 moles of sodium chloride were dissolved to make 1.5 liters of solution, the molarity would be 2.0 mol/1.5 L = 1.3 M.
- **Molality (m)** is the number of moles of solute per kilogram of solvent. For example, if we dissolved 4.0 moles of sugar in 2.0 liters of water (because one liter of pure water weighs exactly one kilogram, these units are interchangeable for molality problems), the molality would be 4.0 mol/2.0 L = 2.0 m.
- **Mole fraction (χ)** is the number of moles of one component in a mixture divided by the total number of moles of all components in the mixture. For example, if we made a mixture by combining 3.0 moles of water and 2.0 moles of isopropanol, the mole fraction of water would be 3.0 mol/(3.0 mol + 2.0 mol) = 0.60.

Dilutions

Dilution is the process by which a solution is made less concentrated by the addition of more solvent. The equation for dilution calculations is $\mathbf{M_1V_1 = M_2V_2}$, where M_1 is the initial molarity of the solution (before dilution), V_1 is the initial volume of the solution, M_2 is the final molarity of the solution (after dilution), and V_2 is the final volume of the solution.

To see how this works, let's figure out how much 0.100 M NaCl we can make from 1.50 L of 0.450 M NaCl solution.

$$M_1V_1 = M_2V_2$$
$$(0.450\ M)(1.50\ L) = (0.100\ M)V_2$$
$$V_2 = 6.75\ L$$

It's as easy and fun as that! Well, it's easy, anyhow.

Factors that Affect Solubility

When making a solution, it's nice to be able to control how much of the solute will dissolve and the speed at which it dissolves. After all, nobody wants to sit around for a day and a half waiting for something to dissolve! Fortunately, chemists have discovered a number of different ways to control both of these factors:

- **Surface area of the solute.** The smaller the particles of solute, the faster they'll dissolve. Because powders have larger surface areas than large crystals, a larger number of the particles being dissolved are exposed to the solvent at any given time, making the solvation process faster. That's why table salt comes in small grains and not big blocks. It's important to note that the surface area affects how quickly a solute will dissolve, but not the quantity of solute that will dissolve.

- **Pressure.** If you want a gas to dissolve in a liquid, increase the pressure of the gas. You can see how this works with an ordinary soda bottle. When the bottle is capped, the carbon dioxide gas is dissolved in the soda and no bubbles can be seen. However, when you open the cap the pressure is released and the carbon dioxide bubbles out of the soda because the solubility has drastically decreased. By increasing the pressure of a gas over a solution, you increase both the rate at which it dissolves and the quantity which can be dissolved.
- **Temperature.** Generally, increasing the temperature of a solvent increases the solubility of most solids (how much stuff can dissolve) and increases the rate at which compounds dissolve. This explains why it's easier to dissolve sugar in hot tea than cold tea. Gases usually dissolve less well in hot solvent, which explains why sodas get flat more quickly on hot days than cold ones.

Solids

You already know what a solid is, and in Chapter 3 we even talked about why ionic solids are hard and brittle. Unfortunately, not all solids are ionic, so we've got to explore the many other types of solids. Let's get to it.

Why Are Solids Solid?

When atoms or molecules form a solid, the solid frequently forms a crystal. These crystals are held together by the attraction between cations and anions (for ionic compounds), by covalent bonds, or even by intermolecular forces. However, regardless of the method used to hold the crystals together, all crystals form regular, repeating three-dimensional structures that are referred to as **crystal lattices.**

Because crystals have regular arrangements, we can think of them as being large structures that consist of building blocks that repeat themselves over and over again. The smallest unit of crystals that can be stacked together to re-create the entire crystal is referred to as a **unit cell.** The unit cells of three of the most common types of crystal structures are shown below:

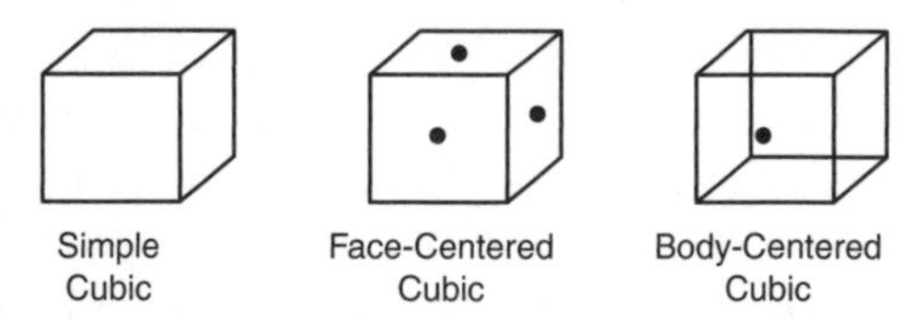

These three crystals represent three very common cubic crystal structures. The name "cubic" in each comes from the shape of the unit cell for each crystal.

Types of Solid

There are five main types of solids, each of which has its own properties and structures. Let's take a look:

- **Ionic solids** (Chapter 3) consist of cations and anions held together by their opposite charges. The crystal structure for each depends mainly on the ratio of the size of the cation to that of the anion.
- **Metallic solids** (Chapter 2) are held together by metallic bonds. Metallic bonds are interesting because in a metallic solid, the positively charged metal nuclei remain in stationary crystalline positions while the valence electrons from each metal are free to wander throughout the solid. As a result of these moving electrons, metals are good conductors of electricity, as well as malleable (able to be pounded into sheets, a measure of how easily it can be deformed) and ductile (able to be made into wires). This theory of metallic bonding is frequently called the **electron sea theory.**

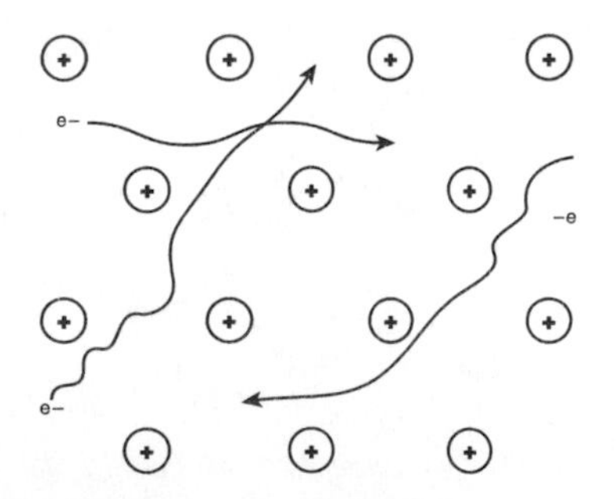

In the electron sea theory, the metal nuclei are locked in place and the electrons float freely throughout the solid.

The Mole Says

Metals frequently have other elements added to them to give them some desired properties such as enhanced hardness or higher melting point. When another element is added to a metal, the resulting compound is called an **alloy.** Common alloys you may be familiar with include stainless steel and 14-carat gold.

- **Network atomic solids** are formed when many atoms are bonded together covalently to form one gigantic molecule. Because all of the atoms in a network atomic solid are held rigidly in place with covalent bonds, network atomic solids have many of the same properties as ionic solids, including extreme hardness, high melting and boiling points, and brittleness. Common examples include quartz, silicon, and diamond.

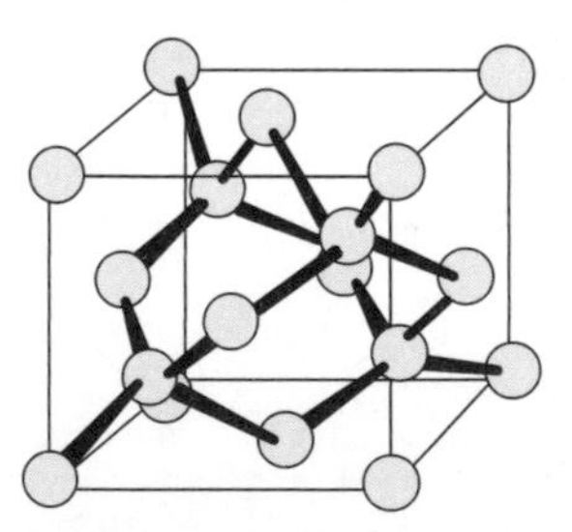

The carbon atoms in a diamond are all held in place by covalent bonds. As a result, diamonds can be thought of as very large covalent molecules.

- **Molecular solids** occur when covalent molecules are held in place by intermolecular forces. Because intermolecular forces are weaker than chemical bonds, molecular crystals have low melting points and are easily broken apart. Examples of molecular solids include sugar and ice.

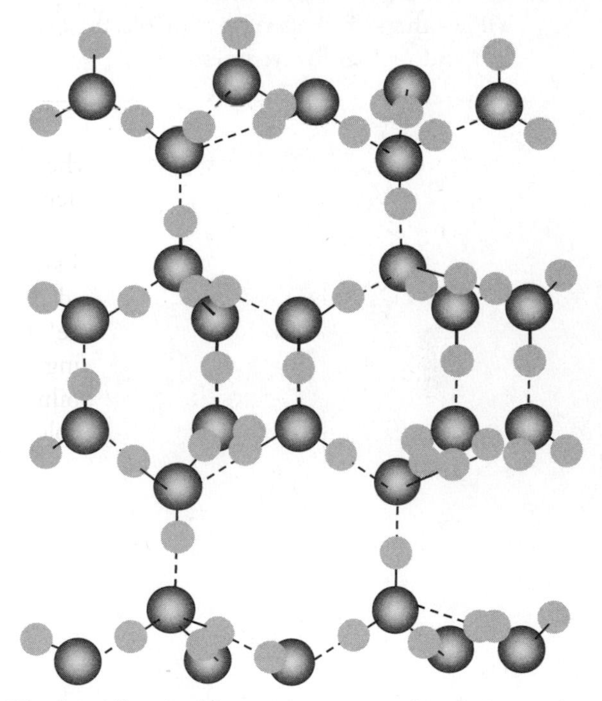

The dotted lines in this structure correspond to the intermolecular forces holding the water molecules together in a molecular solid.

- **Amorphous solids** are materials without any long-term structure. Instead of being arranged into a crystal lattice, the atoms bond in irregular and unpredictable patterns. As a result of this unusual bonding, amorphous solids have a wide range of properties. Some, such as window glass, are hard, brittle, and have a high melting point, while others, such as rubber or plastic, are soft and melt at low temperatures.

Gases

In This Chapter

- The Kinetic Molecular Theory of Gases
- Gas Laws

Gases are harder to conceptualize than liquids or solids. For example, you can drink a liquid and throw a solid at your neighbor's car, but you can't really do much with a gas but breathe it. For this reason, gases were for a very long time the least understood phase of matter.

All that has changed. With the advent of modern chemistry, gases are now well understood. Let's have a look.

What Are Gases?

Gases are the phase of matter in which particles are usually very far from one another, move very quickly, and have weak intermolecular forces. Gases also tend to have the following general properties:

- Gases have low density because the particles are so far apart from each other.
- Gases don't have a fixed shape or volume. If you put a small amount of gas in a room, the gas will expand to fill the entire room.
- Gases are easily compressed because it's easy to push the molecules together so there's less space between them.

Of course, every gas is different, so they'll all have slightly different properties. Because gases are harder to see and handle than solids and liquids, it was difficult for early chemists to understand what they look like on the atomic level. Fortunately, chemists were eventually able to come up with …

The Kinetic Molecular Theory of Gases (KMT)

Gas molecules are always bouncing all over the place at very high speed. As a result, it's hard to study them using standard methods, so chemists had to come up with some new methods that would work more effectively.

Instead of studying individual molecules, chemists used statistical methods to predict what will happen to large numbers of molecules under various sets of conditions. The result is a theoretical model known as the **Kinetic Molecular Theory (KMT).**

Bad Reactions

All theoretical models (including the KMT) only approximate the behavior of the thing being modeled. The approximations that each model uses are designed to make the model easier to use, but because they're not complete, no model is right 100% of the time. This explains why weather models are so lousy at predicting whether it will snow next week.

The following assumptions are used in the KMT to predict the behavior of gases. Again, these assumptions aren't always true, but they do allow us to make some predictions about the behavior of gases:

- **The particles in a gas are infinitely small.** This isn't true—atoms have sizes that can be measured, as do molecules. However, since there's so much space between gas molecules, the actual volume of the molecules themselves makes up only a very small percentage of the total volume.
- **The particles in a gas are in constant random motion.** This assumption is correct—like small children who've eaten too much Halloween candy, gas molecules tend to move in random straight lines until they bump into something else.

- **Gases don't experience intermolecular forces.** Because gas molecules fly around at very high speeds and intermolecular forces are relatively weak, particularly at high temperature and low pressure, it's rare for molecules traveling past each other at high speeds to interact strongly.
- **Gas molecules undergo perfectly elastic collisions.** Elastic collisions are collisions in which kinetic energy is transferred completely from one object to another. If you play pool, you've probably noticed that when one ball hits another in exactly the right way, the first ball stops and the other moves away at the original speed of the first. This, too, is what would happen with molecules if they were to undergo completely elastic collisions.
- **The kinetic energies of gas molecules are directly proportional to their temperatures (in *Kelvin*).** Kinetic energy simply refers to the speed at which gas molecules move around. Not surprisingly, if you add more energy to a gas by heating it up, the molecules will move more quickly.

The Mole Says

Kelvin is the temperature unit that's always used when doing gas law calculations. The reason for this is the preceding assumption—if we used degrees Celsius, then gas molecules would have negative energies at negative temperatures! To convert between degrees Celsius and Kelvin, use the equation **K = °C + 273.**

Why This Is Important: Ideal Gases

We've already said that theories aren't exactly right, and that's true for the Kinetic Molecular Theory as well. Instead of telling us how gases actually behave in the real world, the KMT tells us how gases should behave under perfect conditions. Because no such gas exists, we'll refer to an imaginary "ideal gas."

Why talk about ideal gases at all? Well, because all gases have different intermolecular forces, sizes, shapes, and weights, all gases behave differently in the real world, making it impossible to come up with any law that will fit all of them. Fortunately, the assumptions of KMT are good enough that almost all gases have behavior similar to that of the imaginary ideal gas. As a result, our love of shortcuts causes us to just simplify our calculations by treating real gases as if they were ideal gases after all.

Important Gas Terms and Units

It's time to learn some new terms for working with gases. Because these will come up time and again, make sure you know them!

Pressure

Pressure is defined as the amount of force exerted by the particles in a gas as they hit the sides of the container they're in. To see how this works, have a friend throw tennis balls at you for a few minutes—the force that pushes you back is pressure.

There are several units of pressure that are commonly used to describe gases. **Atmospheres (atm)** are defined as the average atmospheric pressure at sea level, and are commonly used because they're easy to understand. **Millimeters of mercury (mm Hg)** and **torr** are identical to one another; 760 mm Hg or torr equal 1 atm. **Pascals (Pa)** are the metric unit of pressure; there are 101.325 kPa in one atmosphere.

Volume and Temperature

When working with gases, the standard unit for volume is the liter (L) and the standard unit for temperature is Kelvin (K).

Chemistrivia

You may see the term **STP** used when studying gas laws. STP stands for **standard temperature and pressure,** and is equivalent to 1.00 atm and 273 K (or 0° C).

The Gas Laws

Once you understand the kinetic molecular theory, you can begin to make some predictions about how gases will behave under certain conditions. Because chemists are crazy about math, these predictions come in the form of equations for predicting the pressure, volume, or temperature of a gas.

Boyle's Law

What happens to the pressure inside a balloon when you squish it? If you said "It goes up," give yourself a pat on the back! Robert Boyle said the same thing in 1662, but put it in equation form:

$$\mathbf{P_1V_1 = P_2V_2}$$

where P_1 is the pressure of the gas before you squish it, V_1 is the volume of the gas before you squish it, P_2 is the pressure of the gas after you squish it, and V_2 is the volume after squishing.

Let's say that the volume of a balloon is 1.00 L and the pressure inside is 1.00 atm. If you squish the balloon until it's half the volume (0.500 L), the pressure inside, according to Boyle's law, will be:

$$(1.00 \text{ atm})(1.00 \text{ L}) = P_2 (0.500 \text{ L})$$

$$P_2 = 2.00 \text{ atm}$$

It's as simple as that!

Charles's Law

If you take the balloon from the preceding example and place it in the freezer, what will happen to its volume? If you guessed "It decreases," give yourself another pat on the back. The volume of gases increases when you heat them and decreases when you cool them, a rule discovered in 1787 by Jacques Charles:

$$\frac{V_1}{T_1} = \frac{V_2}{T_2}$$

where V_1 is the initial volume of the gas, T_1 is the initial temperature of the gas, V_2 is the volume after the temperature change, and T_2 is the temperature after the temperature change.

Let's imagine that you take a 1.00 L balloon at room temperature (about 25° C) and place it in the freezer (–10° C). The final volume of the balloon would be (after remembering to convert the temperature into Kelvin):

$$\frac{1.00L}{298K} = \frac{V_2}{263K}$$

$$V_2 = 0.883L$$

Gay-Lussac's Law

What would happen if you put a spray can into a hot oven? If you guessed "It will explode," give your already sore back yet another pat! When the gas inside of a spray can gets hot, its pressure increases, which eventually causes the can to blow up. Gay-Lussac said the same thing with his equation:

$$\frac{P_1}{T_1} = \frac{P_2}{T_2}$$

Where P_1 is the pressure before heating the gas, T_1 is the initial temperature, P_2 is the pressure after heating the gas, and T_2 is the temperature after heating it.

The Combined Gas Law

You've probably noticed that Boyle's Law, Charles's Law, and Gay-Lussac's Law all have similar forms. Thankfully, we can save ourselves having to memorize these three laws by memorizing a combination of the three, called the **combined gas law:**

$$\frac{P_1V_1}{T_1} = \frac{P_2V_2}{T_2}$$

Where the subscript "1" refers to the initial pressure, volume, or temperature, and the subscript "2" refers to the final pressure, volume, or temperature.

The advantage this law has over the other three is that it enables us to change more than one variable at a time. For example, if we initially have our 1.00 L balloon in our house where the pressure is

1.00 atm, and the temperature is 298 K, and then release the balloon so it rises to where the pressure is 0.910 atm and the temperature is 215 K, what will the volume of the balloon be?

$$\frac{(1.00atm)(1.00L)}{298K} = \frac{(0.910atm)V_2}{215K}$$

$V_2 = 0.793$ L

The Ideal Gas Law

Because all gases behave in exactly the same way (ideal gases, anyhow), we can write a simple equation to relate the volume of an ideal gas to the number of moles of the gas that's present. This law is called, conveniently enough, the **ideal gas law:**

PV = nRT

where P denotes pressure (in either atm or kPa), V denotes volume (in liters), n is the number of moles of gas, R is the ideal gas constant, and T is the temperature of the gas (in Kelvin). There are two possible values for R: 8.314 L kPa/mol K and 0.08206 L atm/mol K. The value used in each problem will depend on the unit of pressure given. For example, if the pressure of the gas is given in atm, R should be given as 0.08206 L **atm**/mol K.

Example: If a rubber ball has a volume of 10.0 L at a pressure of 1.00 atm and a temperature of 25° C, how many moles of gas can the ball hold?

Solution: P = 1.00 atm, V = 10.0 L, T = 298 K, R = 0.08206 L atm/mol K (because the unit of pressure given was atm). Solving for n, we find that:

(1.00 atm)(10.0 L) = n(0.08206 L atm/mol K) (298 K)

n = 0.409 moles

Notice that we didn't need to specify what gas was in the rubber ball, since all gases would give us the same answer under the kinetic molecular theory!

Dalton's Law of Partial Pressures

Air is a mixture of different gases—for convenience, let's say that it's 78% nitrogen and 22% oxygen. How can we use the gas laws to work with two gases in the same place?

Fortunately, the KMT helps us with this problem. By stating that gas molecules don't interact with one another, we can simply do one calculation for each gas, assuming that the other gas isn't present. As a result, the total pressure of a mixture of gases will be equal to the sum of the partial pressures of all of the individual gases that are in the mixture **(Dalton's law of partial pressures)**:

$$\mathbf{P_{tot} = P_1 + P_2 + P_3 + \ldots}$$

The pressures on the right side of the equation represent the partial pressure of each component gas in the mixture. Partial pressure is defined as the pressure that each gas would exert under the same conditions of temperature and volume if the other gases in the mixture weren't present.

Let's see how this works:

Example: What is the overall pressure of a 50.0 L container that holds 1.00 mol N_2 and 2.00 mol O_2 at a temperature of 298 K?

Solution: To find the overall pressure, we must first find the partial pressure of each gas individually using the ideal gas law:

- The partial pressure of N_2:

  $(P_{nitrogen})(50.0\ L) = (1.00\ mol)(0.08206\ L\ atm/mol\ K)(298\ K)$

  $P_{nitrogen} = 0.489\ atm$

- The partial pressure of O_2:

  $(P_{oxygen})(50.0\ L) = (2.00\ mol)(0.08206\ L\ atm/mol\ K)(298\ K)$

  $P_{oxygen} = 0.978\ atm$

Using Dalton's Law, the total pressure of all the gases in the mixture is:

$$P_{tot} = P_{nitrogen} + P_{oxygen} = 0.489\ atm + 0.978\ atm = 1.467\ atm$$

The Mole Says

You may have noticed in the preceding example that we had to do two different calculations, one for the pressure of nitrogen and one for the pressure of oxygen. However, because ideal gases all behave in exactly the same way (and we assume for simplicity's sake that all gases are ideal), we can simply add the number of moles of gas together and just do one calculation instead of two. Since we have three moles of gas, our calculation then becomes:

$(P_{gas})(50.0 \text{ L}) = (3.00 \text{ mol})(0.08206 \text{ L atm/mol K})(298 \text{ K})$

$P_{gas} = 1.47 \text{ atm}$

Which is exactly the same answer we got before!

Root Mean Square (rms) Velocity

Our last equation involves figuring out the average velocity of the molecules in a gas. This term is called the **root mean square (rms) velocity,** and is calculated using the equation:

$$u_{rms} = \sqrt{\frac{3RT}{M}}$$

where R is always 8.314 J/mol K, T is the temperature of the gas (in K), and M is the molar mass of the gas (in kg/mol—see Chapter 4).

For example, the rms velocity for ammonia (NH_3) at 298 K temperature is found by plugging the temperature and molar mass (0.0170 kg/mol) into this equation with the ideal gas constant:

$$u_{rms} = \sqrt{\frac{3(8.314 J / molK)}{0.0170 kg / mol}} = 661 m / s$$

This may seem counterintuitive. After all, if ammonia molecules traveled 661 meters in a second, wouldn't you be able to smell ammonia instantaneously all over your neighborhood if somebody opened a bottle?

The reason you don't is because of a concept called the **random walk.** If ammonia molecules traveled in a perfectly straight line from your neighbor's house to your nose without bumping into any other gas molecules, you would smell it nearly instantaneously. However, since there are a lot of other gas molecules in the way, it takes a long time for the ammonia molecules to travel all the way to your house. The process by which a gas travels across a room is called **diffusion.**

Chapter 7

Phase Changes

In This Chapter

- Phase changes
- Colligative properties
- Phase diagrams

In the last couple of chapters, we've talked about solids, liquids, and gases. However, we haven't talked at all about how substances change from one phase to the other. With this chapter, all that will change!

Boiling and Condensing

When a substance is heated, the added energy causes the molecules (for a covalent compound) or ions (for an ionic compound) to start moving away from one another. Initially, this causes the liquid to begin evaporating. The pressure caused by the stuff that has evaporated into the gas phase causes the liquid's **vapor pressure,** and not surprisingly, vapor pressure increases as temperature increases because more of the liquid can evaporate. After the vapor pressure of the substance becomes as high as the atmospheric pressure of the surrounding gas, the substance begins to boil. The **normal boiling point** of a liquid is the temperature at which the vapor pressure of the liquid is equal to one atmosphere.

Condensation is the reverse of boiling and occurs when molecules or ions lose enough energy so that the intermolecular forces (for covalent molecules) or electrostatic interactions (for ionic compounds) become strong enough to make the particles stick back together.

The Mole Says

For all substances, the normal boiling point and condensation point are the same, because these processes are simply the reverse of one another. Likewise, the normal freezing point and melting point are the same.

Whenever you do something to a liquid that causes the vapor pressure to decrease, you end up increasing its boiling point. We can see this by comparing pure water to a saltwater solution:

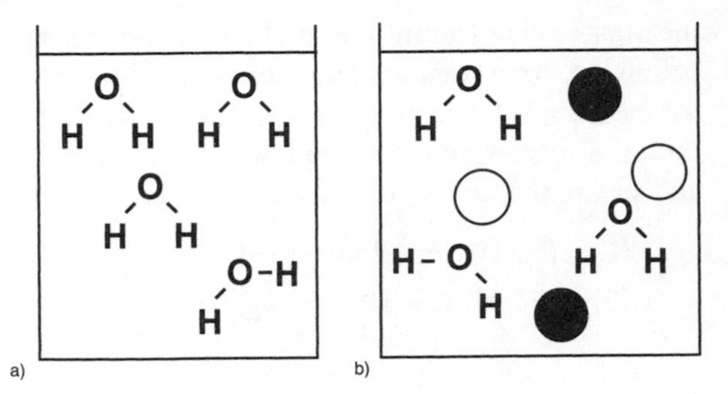

(a)Pure water. (b)Saltwater. The larger circles represent the nonvolatile sodium and chloride ions.

In pure water, any water molecule at the surface of the liquid that gets enough energy to vaporize can do so. This is also true of the saltwater solution, but from the diagram you can see that due to the presence of dissolved sodium chloride, there are fewer water molecules at the surface of the liquid to vaporize. Because the presence of a solute decreases the vapor pressure of the solvent, the solution needs to be heated past the normal boiling point of the pure solvent before it will start to boil.

This increase in boiling point can be calculated using the equation:

$$\Delta T_b = K_b m_{solute}$$

where ΔT_b is equal to the change in the boiling point of the solvent, K_b is the boiling point elevation constant (which is different for every liquid—for water, it's 0.51°C/m), and m_{solute} is the molality of particles of solute in the solution. Molality is a measure of concentration in moles of solute per kg of solvent. As an example, the boiling point of a solution made by adding 2.00 kg of water to 4.00 moles of sugar would be found by solving for ΔT_b and adding that to the normal boiling point of water:

$$\Delta T_b = (0.51° \text{ C/m})(2.00 \text{ m}) = 1.0° \text{ C}$$

$$1.0° \text{ C} + 100° \text{ C} = 101° \text{ C}$$

The Mole Says

When finding the molality for boiling or melting point calculations, make sure to find the molality of the particles of solute, not just the molality of the solute. For example, a 1.00 m NaCl solution would have a molality of 2.00 m for this type of problem because there are 2 ions formed for every NaCl that dissolves. Generally, you should multiply the actual molality by the number of ions for ionic compounds; because the molecules in covalent compounds don't break apart when they dissolve, the actual molality is the same as the molality of the compound itself.

Melting and Freezing

Melting is when energy added to the molecules (in a covalent compound) or ions (in an ionic compound) causes them to break apart from one another so they can move freely among each other. Though the forces that hold the particles together (intermolecular forces for covalent compounds or electrostatic attraction for ions) still keep the particles near one another, the particles have much greater freedom of movement than in a solid. **Freezing** is the opposite of melting, where freely moving particles reattach themselves into a rigid framework via intermolecular forces or electrostatic attraction.

Bad Reactions

Remember, when covalent compounds melt, the forces between molecules are overcome but the covalent bonds within the molecules do not break!

As with boiling, the presence of a solute changes the melting point of a liquid. When a solute is dissolved in a liquid, it's harder for a liquid to freeze (and easier for it to melt). This stems from the makeup of solids—if other particles are present, it makes the crystalline framework of the solid less stable, so it can fall apart at a lower temperature. To determine the melting (or freezing) point of a solution, we use the following equation:

$$\Delta T_f = K_f m_{solute}$$

As with boiling point elevation, ΔT_f refers to the change in freezing point of the solution, K_f is the freezing point depression constant (which is different for every substance, and 1.86° C/m for water), and m_{solute} is the molarity of particles of solute in the solution. For the example in the "Boiling and Condensing" section earlier in this chapter, the freezing point of the sugar solution would be found by subtracting the change in freezing point of the solution from the normal freezing point:

$$\Delta T_f = (1.86° \text{ C/m})(2.00 \text{ m}) = 3.72° \text{ C}$$

$$T_f = 0° \text{ C} - 3.72° \text{ C} = -3.72° \text{ C}$$

The Mole Says

Properties of a solution that depend on its concentration are called colligative properties. Because both the boiling point elevation and freezing point depression depend on the molality of solute present, both are colligative properties. Other colligative properties include vapor pressure reduction, osmotic pressure, color, viscosity (thickness), and taste (not that you should be tasting chemicals).

Sublimation and Deposition

Sublimation is the process by which a solid turns directly into a gas, and **deposition** is the process by which a gas turns directly into a solid. The ice cubes in your refrigerator get smaller over time because the water is subliming and wandering away to elsewhere in your refrigerator, where they are redeposited as solid ice, most often on hot dogs or year-old peas.

Phase Diagrams

So far in this chapter, we've seen how both pressure and temperature affect the state at which matter exists. To put it all together, we use a diagram called a **phase diagram,** which shows the phase of any material under all possible conditions of pressure and temperature. The phase diagram for water is shown below:

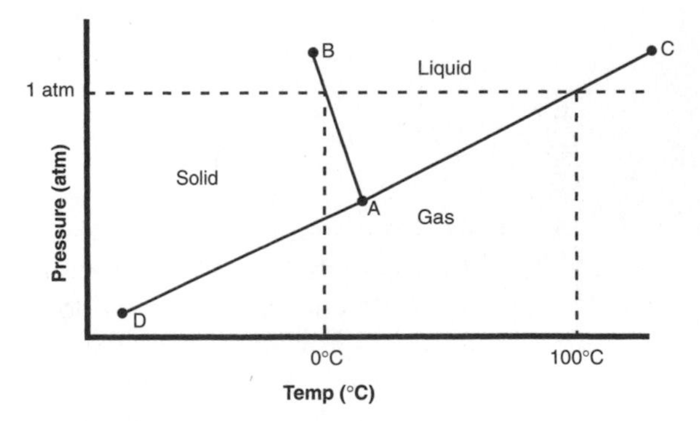

The phase diagram of water.

To read a phase diagram, find the conditions of pressure and temperature that you're interested in for a substance. The region where this point can be found on the graph indicates the stable phase of matter for this substance. For example, from this diagram you can tell that at a pressure of 1 atm and 50° C, water will be a liquid.

Let's use this chart to learn about the important regions of a phase diagram:

- Point A is called the **triple point.** This is the condition of temperature and pressure at which ice, water, and steam are all simultaneously stable. The reason you haven't seen water under these conditions is that it corresponds to pressure much lower than you've probably ever seen in your travels (unless you've been to Mars).
- The line between A and B corresponds to a series of temperature and pressure values where both solid and liquid water are stable. If you follow the dashed line at 1 atm across the chart, you can see that it intersects this line at 0° C, the normal freezing point of water.
- The line between A and C corresponds to the values of pressure and temperature at which the liquid and gas phases are in equilibrium. If you continue following the dashed line, it intersects with line A-C at a temperature of 100° C, the normal boiling point of water.
- The line between A and D represents the values of pressure and temperature at which the solid and gas phases are in equilibrium. By changing the temperature so water moves to the right across this line, you cause it to sublime. By changing the temperature so it moves left across the line, it undergoes deposition.

- Point C is the **critical point.** Past the temperature and pressure of the critical point (called the **critical temperature** and **critical pressure**), the liquid and gas phases of water are indistinguishable from one another and exist in an unusual state called a **supercritical fluid.** Supercritical fluids can be thought of as gases that have been squished down to the point where the molecules are very, very close to one another; as a result, the molecules interact strongly with one another. Because of this, supercritical fluids don't behave like regular gases or like liquids, but may have properties of both.

Phase diagrams are handy because they enable us to figure out what will happen to the state of a material if you change either the pressure or temperature. For example, looking at the phase diagram of water you can see that you can melt ice at 0° C and 1 atm by increasing the pressure. This can be verified by pushing a needle into an ice cube—the increased pressure at the point of the needle melts the ice and the needle slides right in.

Kinetics

In This Chapter

- Energy diagrams
- Factors that affect reaction rates
- Rate laws
- The Arrhenius equation
- Reaction mechanisms

Kinetics is the study of how quickly chemical reactions proceed. You may not know it, but you already have an intuitive understanding of chemical kinetics. For example, you probably understand that it's safe to light a candle with a match, but not safe to put a match in the gas tank of your car, because the rate of the second reaction is much faster than that of the first. If you haven't learned that lesson yet, please take my word for it.

That said, let's learn more!

Energy Diagrams

One of the tools that chemists use to describe how a chemical reaction occurs is called an **energy diagram.** An energy diagram is a graph that shows the amount of energy that the reactants have at all points during the reaction:

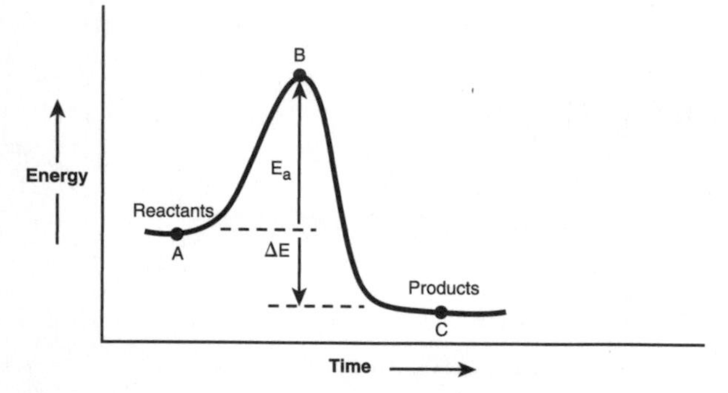

A typical energy diagram for an exothermic reaction.

You can imagine an energy diagram as a map for showing the steps that a reaction must go through before it is finished. At point A, the reactants haven't yet begun to react with one another. When we arrive at point B (called the **transition state** or **activated complex**), the reactants have acquired a quantity of energy called the **activation energy** (**E_a**). The activation energy is the quantity of energy that's required before the reactants can be converted into products—generally, the higher the activation energy, the slower the reaction. The transition state of a chemical reaction is when the reactants have half turned into products; it typically lasts for a *very* short time, with the reactants quickly converting either into products or moving back toward reactants.

Moving down the hill to point C, the products, we see that the reactants lose energy, indicating that the products are more stable than the transition state. Reactions in which the energy difference ΔE between the reactants and products is negative (as in this case) are **exothermic,** meaning they give off energy. Reactions in which ΔE is positive are **endothermic,** meaning they require the addition of energy to take place.

The Mole Says

Imagine a chemical reaction where the activation energy is very low, and the products and reactants have very similar energies. In such a case, the backward reaction from products to reactants may also occur at a measurable rate. Reactions such as this are referred to as being **reversible,** and lead to the establishment of an **equilibrium** (which we'll discuss at great length in Chapter 9). However, you should note that for *any* reaction it's always possible for a reverse reaction to occur, even if it happens at rates too small to measure.

Factors That Affect Reaction Rates

As you might expect, it's frequently handy to know how to control the rates of chemical reactions. After all, you might be unhappy if the gasoline in your car went off explosively or if your steak took 76 hours to cook. The following are some of the factors that can change the rate of a chemical reaction:

- Chemical reactions take place more quickly at high temperatures. We mentioned in the preceding section that the activation energy determines the rate of reaction—by adding more energy, we can more easily overcome the activation energy, increasing the overall rate of the reaction.
- The higher the concentrations of the reactants, the more quickly the reaction will proceed. If you want to perform the reaction $A + B \rightarrow C$, the reaction won't proceed unless molecules of A actually collide with molecules of B. By increasing the concentration of the reactants, you'll increase the number of collisions between A and B, causing an overall increase in reaction rate.
- Increasing the surface area of the reactants increases the rate of reaction. If you have a giant blob of A and a giant blob of B, the reaction $A + B \rightarrow C$ will only proceed at the point where these two blobs actually touch one another. By breaking these blobs into smaller pieces (or better yet, by dissolving them in a solvent), you'll increase the

surface area over which the reactants can touch each other.

- Introducing a **catalyst** increases the rate of a chemical reaction. Catalysts are materials that increase the rates of chemical reactions without themselves being consumed by the reaction. They work by combining the reactants in such a way that the activation energy of the reaction is decreased. Because the activation energy is smaller, the reaction proceeds more quickly.

Chemistrivia

By doing the reverse of these four processes, you can make reactions move less quickly. For example, food spoils less quickly if you keep it in a freezer.

Introduction to Rate Laws

Now that we know from a qualitative standpoint how chemical kinetics works, it's time to roll up our sleeves and get our hands dirty with some equations. To do this, we'll be investigating **rate laws,** which are expressions that show how the rate of a chemical reaction depends on the temperature or concentration of the reactants. For simplicity, we'll assume that all reactions proceed in only the forward direction ($A + B \rightarrow C$), without any of the reverse reaction ($C \rightarrow A + B$) taking place.

Before we get started with what looks like an ugly group of equations, let me give you a suggestion: *Don't panic!* Though the equations may look scary at first glance, they're not all that hard! Even if math makes you break out in big red hives, you'll be able to figure this out if you stay calm!

That said, there are two types of rate laws that we'll be discussing, the first of which is …

Differential Rate Laws

Differential rate laws express how the concentrations of the reactants affect the rate of a chemical reaction. Let's see the general form for a differential rate law, and define the terms:

For a reaction where A + B + … → C, the rate law is:

$$\text{Rate} = k[A]^x[B]^y \ldots$$

This may look confusing, but the terms are easier to follow than you'd expect. [A] and [B] represent the concentration of the reactants, in units of moles per liter (molarity, M). k is the rate constant for the reaction, and indicates the speed at which a particular reaction takes place. x and y are the **orders** of each reactant, and their sum (x+y) is the **reaction order.** Both x and y are determined experimentally, as we'll see in a moment.

Let's find the rate law for the reaction A + B → C, using the following experimental data:

Experiment	[A](M)	[B](M)	Initial rate (M/s)
1	0.0100	0.0100	3.00×10^{-8}
2	0.0200	0.0100	6.00×10^{-8}
3	0.0100	0.0200	1.20×10^{-7}

To find the rate law, we need to find the exponents for the equation:

$$\text{Rate} = k[A]^x[B]^y$$

From our initial data, we can see that the rate doubled between experiment 1 and 2 because we doubled the concentration of compound A. Because the reaction rate is directly proportional to the concentration change of A, the reaction is first-order in A. Likewise, the reaction rate quadrupled between experiment 1 and 3 because we doubled the concentration of B. Because the reaction rate is increased by the square of the concentration change of B, the reaction is second-order in B. The rate equation for this process is shown to be:

$$\text{Rate} = k[A][B]^2$$

and the reaction is said to be first-order in A, second-order in B, and third-order overall.

To take this one step further, we can find the rate constant by plugging the values of rate, [A], and [B] into the rate law we just found. From experiment 1, we find that:

$$3.00 \times 10^{-8}\ \text{M/s} = k[0.01\ \text{M}][0.01\ \text{M}]^2$$

$$k = 0.03\ /\text{M}^2\ \text{s}$$

From this value of k, we can determine the rate of reaction for any concentrations of A and B.

Integrated Rate Laws

Integrated rate laws describe how the concentrations of the reactants change over time. Though they are in a different form than differential rate laws, they're really the same thing with a few mathematical manipulations made for convenience. Let's examine a few common types of integrated rate laws.

First-Order Integrated Rate Laws

Let's say that we've already found that the reaction $A \rightarrow B$ has the rate law:

$$\text{Rate} = k[A]$$

This is a standard differential rate law, and doesn't mention time at all. To introduce a term for time, we need to do some fancy manipulations using calculus (which we won't go into, to preserve our sanity) to find that for a first-order reaction:

$$\ln[A] = -kt + \ln[A_o]$$

where k is the rate constant for the reaction, t is the amount of time since the start of the reaction (in seconds), $[A_o]$ is the initial concentration of reactant A, and [A] is the concentration of reactant A after t seconds have elapsed.

Chemistrivia

For those of you interested in such things, the term **ln** stands for **natural logarithm**. The term "ln[A]" means "the power of the mathematical constant e that generates [A]." For those of you not interested in such things, you can calculate ln[A] by using the handy "ln" button on your calculator.

This equation is handy for the following reasons:

- It allows us to determine how the concentration of our reactant changes as the reaction proceeds, as long as we know what k, t, and $[A_o]$ are.
- The equation is in the form y = mx + b, which is the equation for a straight line. In case your math is rusty, this means that when you make a graph of y (which is ln[A]) versus x (time), for a first order reaction, the slope of the line will be the negative of the rate constant (–k), and the y-intercept (the point where the line crosses over the y-axis) is the $\ln[A_o]$.

One idea often used in kinetics is **half-life,** abbreviated as $\mathbf{t_{1/2}}$. The half-life of a reaction is defined as the amount of time it takes for half of the reactants to be converted into products. For example, if it takes 175 seconds for the concentration of compound A to go from 1.0 M to 0.50 M, the half-life of the reaction is 175 seconds. Likewise, after another 175 seconds has elapsed, the concentration of A will be halved again, to 0.25 M.

We can determine the half-life of a first-order reaction using the equation $\ln[A] = -kt + \ln[A_o]$. Because at $t_{1/2}$ the concentration of A will be half that of its original concentration, we can replace [A] with $[A_o]/2$. If we do a lot of rearranging, we find that:

$$t_{1/2} = \frac{0.693}{k}$$

Second-Order Integrated Rate Laws

Let's consider the rate law for the reaction $A \rightarrow B$, where the reaction is second-order with respect to the concentration of A:

$$\text{Rate} = k[A]^2$$

Using the miracle of calculus to determine how the concentration of A changes over time, we find that:

$$\frac{1}{[A]} = kt + \frac{1}{[A_o]}$$

where the terms are the same as they were for the first-order reaction we discussed in the previous section.

Like the first-order equation, this equation also can be plotted as a straight line of the form $y = mx + b$, except in this case a graph of $1/[A]$ on the y-axis versus time on the x-axis will yield a line with a slope equal to the rate constant and a y-intercept equal to $1/[A_o]$.

This leads us to a handy way to tell whether the process $A \rightarrow B$ is first-order or second-order in A, simply by making two graphs of the available data:

- The first graph is a plot of ln[A] vs. t. If this results in a straight line, the process is first-order in [A].
- The second graph is a plot of 1/[A] vs. t. If this graph results in a straight line, the process is second-order in [A].

For second-order reactions, the half-life is not constant as it is in first-order reactions. In other words, if the initial half-life is 175 seconds, the second half-life will be something different. Because this is a real pain in the butt, from a mathematical viewpoint, it's usually not something people worry about in a first-year chemistry course.

The Mole Says

Occasionally, you can run into a chemical reaction in which the rate doesn't depend at all on the concentration of reactant. Such a reaction is said to be **zero order,** and the integrated rate law is rate = k.

The Arrhenius Equation

We learned earlier that reaction rates increase as temperature increases. Not surprisingly, we can use the wonders of math to figure out specifically how reaction rate and temperature are related.

The equation we use here is called the **Arrhenius equation,** named after the Swedish chemist who came up with it:

$$k = Ae^{-E_a/RT}$$

where k is the rate constant for the reaction, E_a is the activation energy of the reaction, R is the ideal gas constant, T is the temperature (in Kelvin), A is a constant called the **frequency factor,** and e is a button on your calculator. The frequency factor reflects the probability that the molecules colliding will be oriented properly to undergo a chemical reaction.

Probably the easiest way to find out how fast a reaction will go is to figure out the rate constant of the reaction at a given temperature. To do this, you need to know the activation energy of the reaction (E_a, which remains constant at all temperatures) and the rate constant of the reaction at some given temperature (k_1, at temperature T_1). Once you have all this information, you can plug it into a modified version of the Arrhenius equation:

$$\ln\left(\frac{k_2}{k_1}\right) = \left(\frac{E_a}{R}\right)\left(\frac{1}{T_1} - \frac{1}{T_2}\right)$$

By solving this equation using the ideal gas constant (R, 8.31 J/K·mol), we can determine the rate constant of a reaction at an elevated temperature (k_2, at temperature T_2) given the information mentioned previously. Because it's nice to know how heating a reaction will affect the reaction rate in terms more specific than "it will go faster," this equation turns out to be extremely useful for solving kinetics problems.

Reaction Mechanisms

Reaction mechanisms are the series of chemical steps that results in the overall chemical reaction being studied. For example, the equation $A + B \rightarrow C$ may be a one-step reaction, or it may actually take two steps, where the first is $A \rightarrow D$ and the second is $D + B \rightarrow C$.

Why is this important? Well, let's examine what the rate laws for these two processes would be. For the first, $A + B \rightarrow C$, the rate law would be:

$$\text{Rate} = k[A][B]$$

while the reaction mechanism $A \rightarrow D$ followed by $D + B \rightarrow C$ has one rate law for each step:

Rate = k[A] Step 1, $A \rightarrow D$

Rate = k[B] Step 2, $D + B \rightarrow C$

It may seem complicated that the second mechanism would have two rate laws, but it's really not. As it turns out, the overall rate of a chemical reaction is determined by the slowest step in the reaction mechanism, called the **rate-determining step.**

Because the rate equation for the rate-determining step is the overall rate of the equation, we don't have to worry about having two simultaneous equations to solve. In our example, if $A \rightarrow D$ is the slowest step, the overall rate of the second mechanism would be equal to k[A].

The Mole Says

The reason that the slowest step of a multistep process determines the reaction rate makes more sense if you consider an example from real life, drinking soda. Let's say you want to drink 12 cans of soda. The first step in this process involves opening all 12 cans, and the second step involves drinking all 12 cans. Because the "opening soda" step takes 15 seconds and the "drinking soda" step takes 15 hours, it's safe to say that the rate at which the sodas are finished really only depends on the very slow "drinking soda" step. In chemical reactions, the fast and slow steps usually have even more dramatic rate differences than this, making the slow step solely responsible for the overall reaction rate.

From this point, it's a simple matter of matching the actual experimental data to the possible mechanisms we have previously identified. If we find that the actual rate of the reaction is rate = k[A], then the mechanism $A + B \rightarrow C$ is ruled out, and the evidence for the second mechanism is strengthened.

Chapter 9

Solution Chemistry and Equilibria

In This Chapter

- Equilibrium constants
- Heterogeneous equilibria
- Le Châtlier's Principle

In Chapter 8, we briefly mentioned the idea of reversibility, that reactions can move either in the forward or reverse direction. I then ignored any mention of it in the rest of the chapter, dismissing the reverse reaction as being too slow to worry about.

Surprise! Reversibility is back! For many processes the backward reaction is negligibly slow, but not for others. Processes where the backward reaction is important are called **chemical equilibria,** and that's the whole focus of this chapter!

What's an Equilibrium?

Because you're getting to be quite the chemist, I'll start off with the fancy definition: An **equilibrium** is a dynamic, reversible reaction in which the forward and backward rates of reaction are the same.

Since you're still not 100% comfortable with chemistry, let's explain what that means:

- A **reversible reaction** is a reaction that proceeds at measurable rates in either direction. For example, if the forward reaction is $A \rightarrow B$, a reversible reaction will have a measurable reconversion of $B \rightarrow A$.
- In all reversible reactions, the rates of the forward and backward reactions eventually end up being the same after some time has passed. When the concentrations of A and B stop changing, the system is said to be at equilibrium. Important note: The concentrations of A and B don't have to be the same as each other in an equilibrium, and in most cases are not!
- The "dynamic" portion of our definition accounts for the fact that during an equilibrium the forward reaction and the backward reaction are both occurring, even if the concentration of either doesn't change. This means that for every molecule of A that turns into B, a molecule of B turns back into A.

Because reversible reactions go in either direction, we don't use the single-headed arrow (→) to describe the overall chemical process. Instead, we change to a fancy double-headed arrow (↔) to describe equilibria. As a result, if a reaction can proceed in the direction A → B or B → A, we usually write it as A ↔ B.

Equilibrium Constants

All equilibria have different ratios of products to reactants. Let's imagine we're talking about the equilibrium A ↔ B. If the forward reaction A → B is much faster than B → A, the equilibrium concentration of B will be much higher than A. Likewise, if the reverse reaction B → A is much faster than A → B, the equilibrium concentration of A will be higher than B.

So, how do we figure out how much product and reactant are present in an equilibrium? To figure this out, let's use the generic equilibrium:

$$aA + bB \leftrightarrow cC + dD$$

The equilibrium constant for this process (K_{eq}) can be expressed by the ratio of the products to the reactants, or:

$$K_{eq} = \frac{[C]^c[D]^d}{[A]^a[B]^b}$$

where each of the letters in brackets stands for the concentration of that chemical in mol/L (M) when the reaction has reached equilibrium, and the superscripts stand for the coefficient of that chemical in the equation. For gases, the equilibrium constant is determined in the same way, except that partial pressures in atm are used.

The equilibrium constant is important because it tells us where the equilibrium lies. The larger the equilibrium constant, the further the equilibrium lies toward the products. For example, an equilibrium constant of 10^6 suggests that most of the chemical species are products, while an equilibrium constant of 10^{-6} indicates that nearly all molecules are reactants.

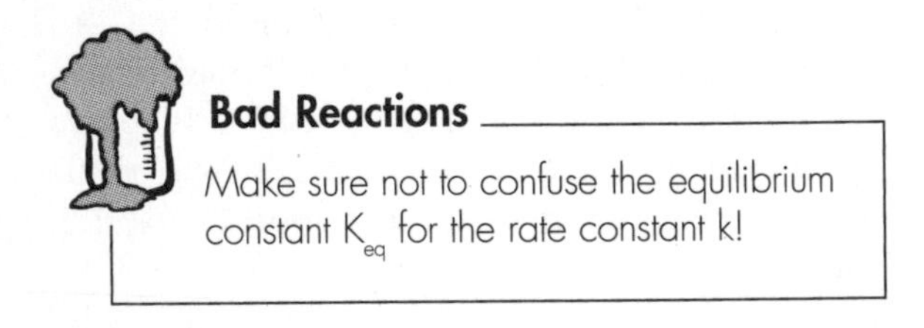

Bad Reactions

Make sure not to confuse the equilibrium constant K_{eq} for the rate constant k!

Let's see an example of how to use an equilibrium constant.

Example: Given the reaction $A \leftrightarrow B + C$, what will the equilibrium concentrations of B and C be if the initial concentration of A is 1.00 M and the equilibrium constant is 1.86×10^{-6}?

Solution: Let's write the equilibrium expression for this process:

$$K_{eq} = \frac{[B][C]}{[A]}$$

From this, we can figure out what the concentrations of each species will be at equilibrium. This is shown in the chart below (if it's not obvious how we got this, don't panic, we'll explain it!):

Species	Initial Molarity	Change	Final Molarity
A	1.00 M	–x M	(1.00 – x) M
B	0 M	+x M	x M
C	0 M	+x M	x M

Let's talk about where these numbers came from:

- The initial concentration of A was given in the problem as 1.00 M. Because some quantity of A is converted to B and C during the reaction, but we don't know how much, we'll refer to the change in the concentration of A as being –x M. As a result, the concentration of A at equilibrium will be (1.00 – x)M.
- Likewise, for every molecule of A that breaks apart, one molecule each of B and C are formed. As a result, the change in the initial concentration of B and C (0 M) is equal to the change in A, making the final concentration of B and C equal to x M.

To figure out what the concentration of each species is, all we have to do is plug the concentrations of A, B, and C back into the equilibrium expression and solve for x. When we do this, we get:

$$1.86x10^{-6} = \frac{[x][x]}{[1.00 - x]}$$

The bad news is that to solve this equation for x, we need to go back to our math classes and use the quadratic equation, which is way too big and complicated to be much fun. The good news is this: If K_{eq} is very small, the quantity of x formed will be very small compared to the original quantity of the reactant. As a result, we can omit the x in the denominator to simplify the [1.00 – x] term because [1.00 – x] will be roughly equal to 1.00. Our new (and much easier) expression to solve is:

$$1.86x10^{-6} = \frac{[x][x]}{[1.00]}$$

$$x = 1.36 \times 10^{-3} \text{ M}$$

The final concentration of B and C is 1.36×10^{-3} M, and the final concentration of A is [1.00 – 0.00136 M] which is roughly equal to 1.00 M. Incidentially, if you do end up using the quadratic equation (which is necessary in cases with a moderate K_{eq} but not covered here), remember to reject a solution that would result in a negative pressure. In this example, the math is simple enough that you can avoid the quadratic equation and solve by taking the square root of both sides.

Heterogeneous Equilibria

So far, all the equilibria we've talked about have had chemicals in the same phase. In solutions, they're dissolved, and in gaseous equilibria, they're all gases. Equilibria where all species are in the same phase are referred to as **homogeneous equilibria.**

Likewise, if all the species are not in the same phase, the equilibrium is referred to as being a **heterogeneous equilibrium.** An example of a heterogeneous equilibrium is when an ionic compound partially dissolves (or in chemical terms, dissociates) in water; the undissolved ionic compound is a solid, and the dissolved portion is aqueous.

The main way that the rate law in a heterogeneous equilibria differs from that of a homogeneous equilibrium is that the pure substance, whether it's a solid or a liquid, is left out of the expression. We'll see how this works in the example below, where the dissociation of $CaCO_3$ in water illustrates this point:

$$CaCO_{3(s)} \leftrightarrow Ca^{+2}{}_{(aq)} + CO_3{}^{-2}{}_{(aq)}$$

You'd expect the equilibrium expression for this process to be:

$$K_{sp} = \frac{[Ca^{+2}][CO_3{}^{-2}]}{[CaCO_3]}$$

However, because calcium carbonate is a pure solid, its concentration relative to itself is one. As a result, we leave it out of the equilibrium expression, giving us:

$$K_{sp} = [Ca^{+2}][CO_3{}^{-2}]$$

The Mole Says

The "sp" after the K term in solubility problems stands for **solubility product,** giving K_{sp} the term **solubility product constant.** Generally, what you'll see is that whenever you have an equilibrium, you'll have a K term as an equilibrium constant along with a subscript below it indicating the type of equilibrium. Some examples: K_{eq} is an equilibrium constant for a generic equilibrium; K_{sp} is the equilibrium constant for when something dissolves in water; K_a is the equilibrium constant for when an acid dissociates in water; K_p is the equilibrium constant for a gaseous equilibrium (the "p" stands for "pressure"); and so on.

Example: What's the concentration of Ca^{+2} ions in a saturated $CaCO_3$ solution?

$$K_{sp}(CaCO_3) = 4.5 \times 10^{-9}.$$

Solution: To solve this, we need only plug the K_{sp} value for calcium carbonate into the solubility product expression for $CaCO_3$. Because we don't know the concentration of Ca^{+2} or CO_3^{-2}, we plug in "x" for both of them.

$$K_{sp} = 4.5 \times 10^{-9} = [x][x]$$
$$x = 6.7 \times 10^{-5}M$$

Le Châtlier's Principle

So far, we've pretended that once an equilibrium is established, we leave it alone. This isn't always true. For example, it's usually handy to push an equilibrium so the quantity of product formed will be greater.

The overall rule we use to predict how equilibria will behave when the conditions are changed is called **Le Châtlier's Principle,** which states the following: If you change the conditions of an equilibrium, the equilibrium will shift in a way that tends to compensate for whatever you did in the first place.

Let's see how this works when we do a variety of interesting things to an equilibrium:

- When you change the concentration of one of the chemical species in an equilibrium, the equilibrium will shift to reduce the quantity of that chemical. For example, if we have an equilibrium $A + B \leftrightarrow C$ and we add a bunch of extra B to the equilibrium, the equilibrium will shift to the right (toward the products) to make more C. By doing this, the quantity of B you added will be decreased. Likewise, if you were to remove some of A, the equilibrium would shift to the left (toward the reactants) to increase the amount of A present.
- If you have a gaseous equilibrium, you can shift the position of the equilibrium by

squishing or compressing the mixture. For example, imagine the gaseous equilibrium $A_{(g)} + B_{(g)} \leftrightarrow C_{(g)}$. If you squish the mixture into a smaller area, the partial pressures of all the gases will increase. In response to this, the equilibrium will shift to decrease the overall pressure of the system. Because there are fewer moles of gas to the right of the equation, the equilibrium will shift toward the products. This is how ammonia is made!

- Changing the temperature can change the position of an equilibrium. For example, let's consider an exothermic reaction: $A \leftrightarrow B$ + energy. If we increase the temperature, and hence the energy, it's as if we're adding to the products side of the equation. As a result, the equilibrium will shift toward A to decrease the amount of energy present. Likewise, if we reduced the temperature, the equilibrium will shift to the right to increase the amount of energy present.

Chapter 

Acids and Bases

In This Chapter

- The three definitions of acids and bases
- Determining pH
- Titrations
- Buffers

Many of the things we see in our everyday lives are either acids or bases. From the acids in batteries to the bases in soaps, acid-base chemistry plays a large role in our lives. Let's learn more!

What Are Acids and Bases?

This is an excellent question, and unfortunately, there are three main definitions of acids and bases that you need to know about: Arrhenius acids and bases, Brønsted-Lowry acids and bases, and Lewis acids and bases. On the positive side, these three definitions are closely related to one another.

Arrhenius Acids and Bases

In the late 1800's, Svante Arrhenius gave the following definitions of acids and bases:

- **Arrhenius acids** are compounds that give off H^+ ions, which are protons in water. These ions are usually called hydronium ions, but may also be called protons and may be given the symbol H_3O^+, which is what is formed when H^+ combines with H_2O. The formulas of acids usually have H as the first letter (as in hydrochloric acid, HCl; nitric acid, HNO_3; and sulfuric acid, H_2SO_4). Common exceptions are organic carboxylic acids, which end with -COOH, as in acetic acid (CH_3COOH).
- **Arrhenius bases** are compounds that give off OH^- (hydroxide) ions in water. According to this definition, hydroxides are bases.

Brønsted-Lowry Acids and Bases

In the early 1900's, an alternate definition of acids and bases was proposed to account for the fact that ammonia can neutralize the acidity of hydrochloric acid, even though it's not a hydroxide. According to this new definition:

- **Brønsted-Lowry acids** donate hydronium ions to another compound. For example, if HCl gives a hydronium ion to another compound, it's a Brønsted-Lowry acid. This is very similar to the Arrhenius definition.

- **Brønsted-Lowry bases** accept hydronium ions from other compounds. For example, if ammonia (NH_3) accepts a hydronium ion, it is a base.

We can see how this works by writing an equation for the reaction of hydrochloric acid with ammonia:

$$HCl + NH_3 \rightarrow NH_4^{+1} + Cl^{-1}$$

You can see from this definition that hydrochloric acid is the Brønsted-Lowry acid and ammonia is the base, because hydrochloric acid gives H^+ to ammonia.

One common idea we use with Brønsted-Lowry acids and bases is that of **conjugate acids and bases.** If you take a look at the products of the reaction above, you see that the ammonium ion is capable of giving a hydronium ion to the chloride ion, which is capable of receiving it. As a result, the ammonium ion on the right is a Brønsted-Lowry acid and the chloride ion is a Brønsted-Lowry base. Because the ammonium ion was formed from the basic ammonia molecule on the left, the ammonium ion is said to be the **conjugate acid** of ammonia, and the two together are referred to as a **conjugate acid-base pair.** Likewise, another conjugate acid-base pair in this equation consists of hydrochloric acid and its conjugate base, Cl^-.

Lewis Acids and Bases

In the Brønsted-Lowry definition, a base is a compound that accepts a hydronium ion from an acid. Brønsted-Lowry bases all have lone pairs of electrons that they use to grab these hydronium ions from acids, as shown below:

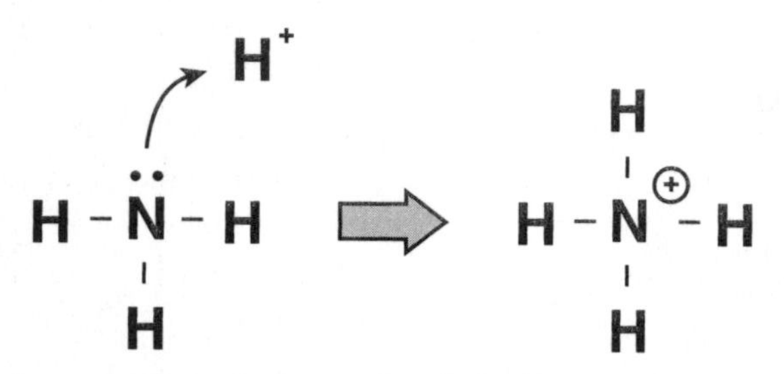

Ammonia can grab a proton from hydrochloric acid with its lone pair electrons.

Extending this somewhat, we get a new definition of acids and bases, called **Lewis acidity and basicity:**

- **Lewis acids** are compounds that accept an electron pair from another compound. Because HCl was accepting a lone pair from ammonia, it was the Lewis acid in this example.
- **Lewis bases** are compounds that donate an electron pair to another compound. In our example above, ammonia used its lone pair electrons to grab onto the H on hydrochloric acid.

Lewis acidity and basicity is a much broader definition than the others we've learned because it doesn't confine itself to the movement of H^+ or OH^- ions. The following shows an example of a Lewis acid-base reaction in which a bond forms between two compounds.

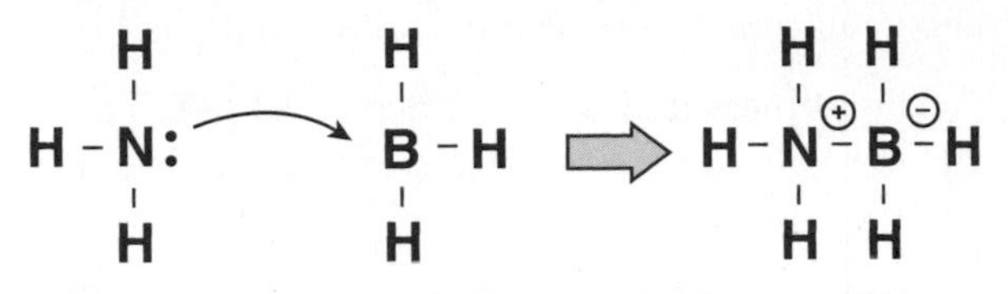

Ammonia, a Lewis base, uses its lone pair electrons to form a bond with BH_3, a Lewis acid.

Properties of Acids and Bases

It's frequently possible to tell acids and bases apart from one another by some of their easily observed chemical and physical properties. While I certainly wouldn't suggest you taste or touch acids or bases in the laboratory, the table below indicates how you might tell household acids from bases:

Property	Acid	Base
Taste	Sour (vinegar)	Bitter (baking soda)
Smell	Burns nose	No smell (except NH_3!)
Texture	Sticky	Slippery
Reacts with …	Metals to form H_2,	Oils, and fats

The pH Scale

As you might imagine, it's useful for chemists to tell how acidic a compound is. For example, I finished an acidic soda while writing the previous section of this chapter. Not surprisingly, I would be very unhappy if the maker of the soda made it a thousand times more acidic than it should be!

Acidity is measured by a number called **pH.** The pH of a solution can be determined using the equation:

$$\mathbf{pH = -\log[H^+]}$$

where $[H^+]$ is the molarity of H^+ ions present. The value of pH itself has no unit, so don't add M or pH after the pH value.

Solutions with a pH of exactly 7.00 are considered neutral. If a solution has a pH less than 7.00 it is an acid, and if it's greater than 7.00 it is a base.

The Mole Says

Though a solution with a pH of 7.001 is basic from a chemical standpoint, it's common for chemists to refer to solutions with pH values near 7 as neutral. Many of the things you consider neutral are actually very slightly acidic or basic.

Finding pH Using Indicators

Indicators are compounds that turn different colors when exposed to acids or bases. For example, litmus, a common indicator, is red in acidic solutions and blue in basic solutions. Phenolphthalein is colorless in acidic solutions and pink in basic solutions. Though these indicators may not tell you the exact pH of a solution, they can give you a rough idea of whether something is acidic or basic.

Finding the pH of a Strong Acid

Strong acids are acids that completely dissociate (break apart into H^+ and an anion) when placed in water. Common examples include HNO_3, H_2SO_4, HI, HCl, and $HClO_4$. Because these acids break apart so well, the concentration of H^+ in the solution is the same as the original concentration of the acid that's present. As a result, the pH of a 0.00500 M HCl solution would be found using the pH equation, plugging 0.00500 M in as the value of $[H^+]$:

$$pH = -\log(0.00500\ M) = 2.30$$

Finding the pH of a Weak Acid

Weak acids are acids that dissociate only slightly in water. As a result, their dissociation is an equilibrium with the following general form:

$$HA \leftrightarrow H^+ + A^-$$

and the equilibrium expression:

$$K_a = \frac{[H^+][A^-]}{[HA]}$$

Because the concentration of H^+ is not the same as the original concentration of the acid, we need to use the preceding equilibrium expression to determine the pH of a weakly acidic solution, the same way we did in Chapter 9. Let's see an example:

Example: What's the pH of a 0.500 M acetic acid solution?

$K_a(CH_3COOH) = 1.75 \times 10^{-5}$

Solution: The equation for acetic acid dissolving in water is $CH_3COOH \leftrightarrow CH_3COO^{-1} + H^+$, which yields the equilibrium expression:

$$K_a = \frac{[CH_3COO^{-1}][H^+]}{[CH_3COOH]}$$

Initially, the concentration of acetic acid is 0.500 M. However, once it has partially dissociated, the concentration decreases by an unknown amount, which we'll denote "x." Likewise, the concentration of the acetate and hydronium ions when we place acetic acid in water will rise by the same amount "x." As a result, the concentration of acetic acid at equilibrium is (0.500 – x) M and the concentration of both hydronium and acetate ions are x M. Plugging this into the equilibrium expression, we get:

$$1.75x10^{-5} = \frac{[xM][xM]}{[(0.500 - x)M]}$$

As in Chapter 9, we'll assume that x is very small when compared to 0.500 M, so we can simplify this equation as:

$$1.75x10^{-5} = \frac{[xM][xM]}{[0.500M]}$$

$$x = 0.00296 \text{ M}$$

Because x is equal to the concentration of hydronium ions present, we need only plug this value into the expression for pH to find the acidity of the solution:

$$pH = -\log[H^+]$$

$$pH = -\log(0.00296 \text{ M})$$

$$pH = 2.53$$

Finding the pH of Basic Solutions

Let's find the pH of a 0.0500 NaOH solution. To do this, we simply plug the value 0.0500 into the equation …

Well, we have a problem. NaOH isn't an acid, and all of the equations we've worked with so far require the concentration of the hydronium ion, H^+. What do we do when we have a base?

As it turns out, there are H^+ ions present even in basic solutions. This happens because water itself breaks up (a process called **autoionization**) to form H^+ and OH^- ions:

$$H_2O \leftrightarrow H^+ + OH^-$$

So wherever water is present (as in any aqueous solution) there is a small amount of both H^+ and OH^- ions. The equilibrium constant for the autoionization of water has the symbol K_w, and is equal to 10^{-14}. This expression is shown below:

$$K_w = [H^+][OH^-] = 10^{-14}$$

Because this equation gives us an easy way to relate the quantity of H^+ to the quantity of OH^- in aqueous solutions, we can now find the pH of bases as well as acids.

Back to our problem: We have a base concentration, but we can't plug that into the equation for pH. Therefore, we must first convert the concentration of base to the quantity of acid present using the equilibrium expression for the autoionization of water:

$$K_w = [H^+][OH^-] = 10^{-14}$$

$$[H^+][0.0500\ M] = 10^{-14}$$

$$[H^+] = 2.00 \times 10^{-13}\ M$$

Now that we have the concentration of acid, all we have to do is solve for the pH of the solution:

$$pH = -\log[H^+] = -\log[2.00 \times 10^{-13}\ M]$$

$$pH = 12.7$$

This is a very basic solution, as we might expect.

Titrations

I found an unlabeled bottle in the stockroom of my lab one time. It had no label, but from the scorching smell I could tell it was nitric acid. Unfortunately, I couldn't do much with it unless I knew the acid concentration.

Fortunately, I remembered that I knew a method for figuring out the concentration of an acid (or base, for that matter) called **titration.** Titration is a method in which neutralization reactions are used to determine the concentration of either an acidic or basic solution.

Here's how it works:

- In an acid-base reaction, acids and bases neutralize one another. I can neutralize the acid in the bottle with a base, eventually turning it into a neutral solution.
- When the entire solution is neutral, the number of moles of base that I added (which I can measure exactly) will be equal to the number of moles of acid that was originally present (which I'm trying to find). The point when this happens is called the **equivalence point.**
- If I add an indicator to the acid before adding base, a color change will tell me when the solution becomes neutral (this point is called the **endpoint,** which isn't exactly equal to the equivalence point because the indicator doesn't change at a pH of exactly 7.0000).

Another way of expressing that the number of moles of the base will be the same as the number of moles of acid originally present is with the equation:

$$\mathbf{M_aV_a = M_bV_b}$$

where M_a represents the molarity of the acid, V_a is the volume of the acid, M_b is the molarity of the base, and V_b is the volume of the base.

Using all of this knowledge, I set up an experiment where I placed 175 mL of nitric acid into a beaker and added 1.00 M NaOH to it until it was completely neutralized. This required the addition of 365 mL of NaOH solution. Using the preceding equation, where M_a is unknown, V_a is 175 mL, M_b is 1.00 mL, and V_b is 365 mL, I found that the original concentration of the acid is:

$$M_a\ (175\text{ mL}) = (1.00\text{ M})(365\text{ mL})$$

$$M_a = 2.09\text{ M}$$

Why anybody would want a 2.09 M nitric acid solution is still unknown to me.

Buffers

Writers love to drink soda when writing books. In fact, today I've consumed 11 cans of diet cola. Because diet cola is fairly acidic, you might expect that the pH in my body would become much lower than the normal near-neutral condition, and I would slip into a coma and die.

Surprise! I'm not dead! Thankfully, blood, like many solutions, is a buffer. **Buffers** are solutions that consist of a weak acid and its conjugate base and resist changes in pH when either acid or base is added to them.

Let's consider a buffered solution containing acetic acid as the weak acid and sodium acetate as its conjugate base. If we were to add hydrochloric acid (a very strong acid) to this solution, the following reaction would occur:

$$HCl + NaCH_3COO \rightarrow NaCl + CH_3COOH$$

There's still just as much acid here as before. The difference is that the strong acid, HCl, has been converted into a weak acid, CH_3COOH. Because CH_3COOH changes the pH of a solution much less than HCl does, the overall pH of the solution doesn't change much.

Likewise, if we were to add NaOH, a strong base, to the solution, we'd have the following reaction:

$$NaOH + CH_3COOH \rightarrow NaCH_3COO + H_2O$$

where sodium hydroxide, a very strong base, is converted into sodium acetate, a very weak one. Because a weak base changes the pH of a solution much less than a strong one, the pH stays relatively stable.

We can determine the pH of a buffered solution using the **Henderson-Hasselbalch equation:**

$$pH = -\log K_a + \log \frac{[base]}{[acid]}$$

where K_a is the acid dissociation constant of the weak acid in the buffered solution. If you know the concentration of the weak acid and its conjugate base, you can easily find pH using this equation.

Chapter 11

Organic Chemistry

In This Chapter

- Hydrocarbons
- Isomerism
- Functional groups
- Basic organic reactions

In my opinion, organic chemistry is the most interesting branch of chemistry. Unfortunately, any first-year chemistry course is unlikely to convince you of this. Because organic chemistry is such an enormous topic, instructors usually just give students a great big list of "simple" things to memorize. While this chapter may not be the most exciting one in this book, and definitely doesn't do a good job of showing how much fun organic chemistry is, it will tell you what you need to know to make your instructor happy.

What Is Organic Chemistry?

Organic chemistry is the study of carbon-containing compounds. Organic molecules contain carbon and hydrogen, and frequently other elements such as oxygen, nitrogen, the halogens, sulfur, phosphorus, and just about anything else you can think of. The only carbon-containing compounds that don't qualify as organic are CO, CO_2, and compounds containing the carbonate ion (CO_3^{-2}).

Hydrocarbons

Hydrocarbons are molecules containing only carbon and hydrogen. However, because carbon and hydrogen can be combined in so many different ways, there are still lots of ways to put them together.

Alkanes

Alkanes (also called **saturated hydrocarbons**) are hydrocarbons that contain only single bonds. The naming system for alkanes is based on the names of alkanes in which all of the carbon atoms are arranged in straight chains. The first eight alkanes are shown in the next figure:

Number of carbon atoms	Name	Formula	Structure
1	methane	CH_4	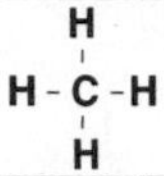 H H - C - H H
2	ethane	C_2H_6	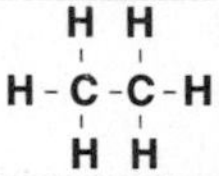 H H H - C - C - H H H
3	propane	C_3H_8	
4	butane	C_4H_{10}	
5	pentane	C_5H_{12}	
6	hexane	C_6H_{14}	
7	heptane	C_7H_{16}	
8	octane	C_8H_{18}	

The first eight straight-chain hydrocarbons. If you remember only one thing from this chapter, remember this!

The Mole Says

In this figure, all of the atoms are shown for methane and ethane, but only straight lines are drawn for the other molecules. The common shorthand for drawing organic molecules is to assume that the intersections and ends of all lines correspond to a carbon atom, and that hydrogen atoms are added to these carbon atoms until all the carbon atoms have four bonds. By this method, you could show methane as a single dot and ethane as a straight line (though that is *never* done).

To name alkanes, follow these rules:

- The name of an alkane is based on the largest unbroken chain of carbon atoms in the molecule. For example, the following figure shows a **hexane,** because the longest carbon chain has six atoms:

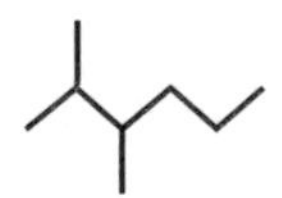

One of the hexane family of molecules.

- Any group that hangs off the longest chain is named based on the number of carbon atoms it contains, except that the ending "-yl" replaces "-ane." In our example, each of the groups hanging off of the hexane chain has one carbon—because alkanes with one carbon are named methane, we refer to these as **methyl groups.**

- Because there are many positions where a group can be attached on the chain, we have to indicate which carbon atom it is attached to. To do this, we number the carbon atoms from each end of the chain such that the side chains (called alkyl groups if they contain only carbon and hydrogen) have the smallest possible numbers. In our earlier example, we can number the chain in two possible ways, as shown here:

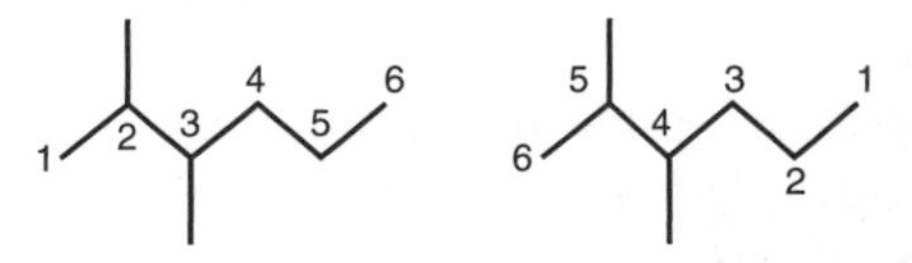

The numbering scheme on the left is correct because it assigns to the two methyl groups the smallest possible numbers.

- If there is more than one of a certain kind of group connected to the main carbon chain (called substituents), use the prefixes "di-" to indicate there are two, "bi-" to indicate three, "tetra-" to indicate four, and so on to indicate how many there are. Before these prefixes, indicate the carbon atom in the chain that each substituent is stuck to. In our example, the molecule is "2,3-dimethylhexane," indicating that it consists of a six-carbon main chain with methyl groups at the 2- and 3- positions.

- If there is more than one different type of substituent, put them in alphabetical order. For example, the molecule below is "3-ethyl-2,4-dimethyloctane":

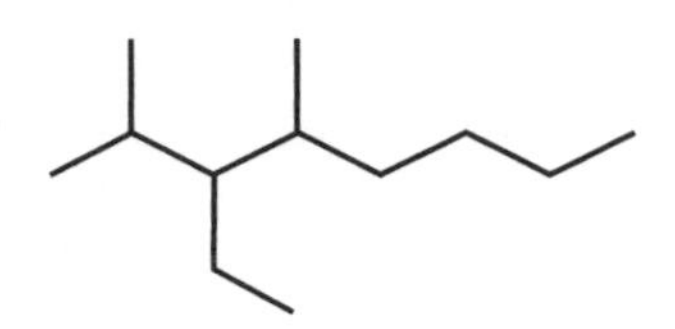

3-ethyl-2,4-dimethyloctane.

Alkenes and Alkynes

Alkenes contain at least one C-C double bond and **alkynes** contain at least one C-C triple bond. Because additional hydrogen atoms can be added to these molecules, they're referred to as **unsaturated hydrocarbons.**

Alkenes and alkynes are named in much the same way as alkanes, except that the ending of the molecule is "-ene" or "-yne," respectively. Additionally, a number is added before the name of the longest chain to indicate the position of the multiple bond, as seen below:

2-methyl-1-pentene.

Cyclic Hydrocarbons

Carbon atoms frequently form rings. These rings are called **cycloalkanes** if they have only single C-C bonds, **cycloalkenes** if they have at least one C-C double bond, and **cycloalkynes** if they have at least one C-C triple bond. The most common rings are cyclopropyl- and cyclohexyl- rings.

Aromatic Hydrocarbons

Aromatic hydrocarbons are cyclic hydrocarbons with alternating carbon-carbon single and double bonds. Benzene, the best known aromatic hydrocarbon, is unusually stable because the electrons can travel around the entire ring, a situation known as **delocalization.**

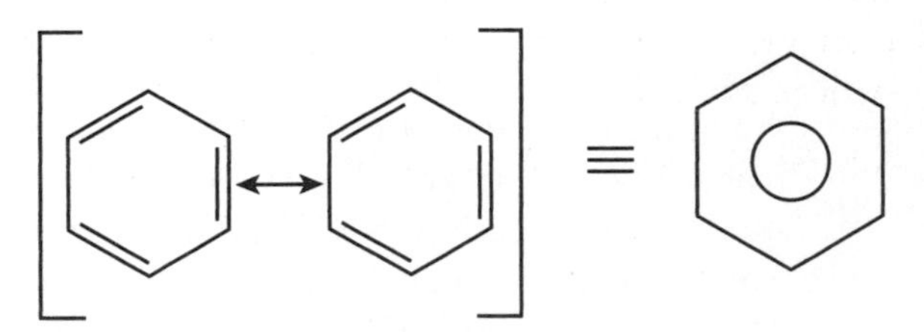

The two resonance structures of benzene on the left show how the electrons in the double bonds can switch positions. The third diagram is often shown to indicate that benzene behaves more stably than most molecules containing three double bonds.

Isomers

Just because two molecules have the same formula doesn't mean they have the same structure.

Molecules that have the same molecular formula but different structures are called **isomers.** There are two different types of isomerism—constitutional isomers (also called structural isomers) and stereoisomers.

- **Constitutional isomers** are molecules that have the same formulas but differ in the order that the atoms are connected to one another. Examples of constitutional isomers are shown below:

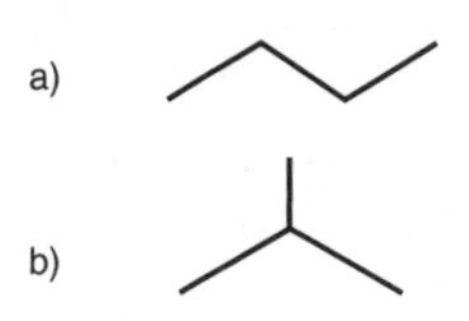

Constitutional isomers butane and 2-methylpropane.

- **Stereoisomers** are molecules with atoms bonded in the same order, but with different spatial orientations. Your two feet have the same relationship to one another that stereoisomers do—they have the same pattern of five toes (I hope!) but are different in that they don't fit as well into your left shoe.

Functional Groups

As mentioned earlier, there are a lot of different atoms that can be present in an organic molecule. The method they are incorporated into an organic molecule is referred to as their **functional group.**

The following figure shows the most common organic functional groups you're likely to see (and should *definitely* memorize). Keep in mind, anytime you see "R" in an organic diagram, that's just a generic way of saying generic organic group—for example, R-Br can be CH_3Br, C_6H_5Br, or just about anything else organic you can imagine with –Br on it.

Functional group	General Formula	Example	Name
alkyl halide	R-X	CH_3Br	bromomethane (methyl bromide)
alcohol	R-OH	CH_3OH	methanol
ether	R-O-R' (R groups not necessarily the same)	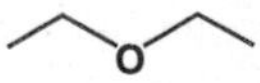 O	diethyl ether
aldehyde	O ‖ R—C—H	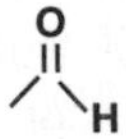 O ‖ H	ethanal (acetaldehyde)
ketone	O ‖ R—C—R'	 O ‖	propanone (acetone, dimethyl ketone)
carboxylic acid	O ‖ R—C—OH	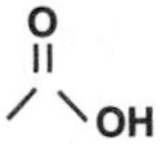 O ‖ OH	ethanoic acid (acetic acid)
ester	O ‖ R—C—O—R'	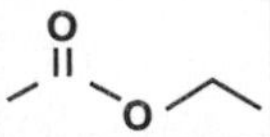 O ‖ O	ethyl ethanoate (ethyl acetate)
amine	R—N—R'' | R'	N	triethylamine

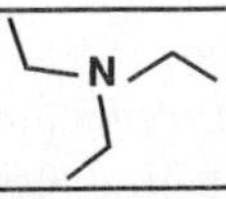

The most common organic functional groups and their names.

Chemistrivia

It's not uncommon for organic chemists to be able to tell the difference between different classes of molecules by their smells. For example, esters usually have pleasant floral or fruity smells, and amines usually smell very bad.

Organic Reactions

There are a lot of different types of organic reactions. For example, the best organic chemistry text I have has 1,495 pages of reactions. Because you don't want to read a huge textbook, I'll try to shorten it to a couple of pages.

- **Addition reactions** occur when an alkene or alkyne becomes saturated with atoms of other elements. Common types of addition reactions include **hydrogenation** (where the double bond is replaced with hydrogen atoms) and **halogenation** (where the double bond is replaced with halogen atoms, or one hydrogen and one halogen atom).
- **Free-radical substitutions** occur when alkanes react with halogens in the presence of light, to form alkyl halides. The mechanism has three steps: In the first (**initiation**), the halogen molecule is broken apart into atoms with unpaired electrons (called **free radicals**) by light; in the second (**propagation**), the

halogen radical forms alkane radicals, which then react with halogen molecules to form alkyl halides and additional halogen radicals; in the third (**termination**), free radicals combine with one another.

- **Oxidation** occurs when alcohols are converted to aldehydes, ketones, or carboxylic acids.
- **Condensation reactions** take place when two molecules combine with one another in a way that also results in the formation of water. The example below shows two molecules of methanol reacting to form dimethyl ether and water:

$$CH_3OH + HOCH_3 \longrightarrow H_2O + CH_3\text{-}O\text{-}CH_3$$

The condensation reaction of two molecules of methanol to form dimethyl ether and water.

- **Polymerization reactions** are when very small molecules link up with one another to form much larger chains of molecules called **polymers.** Most of the plastics that you're familiar with, such as Teflon and polyethylene, are polymers. Polymerization reactions are frequently also free-radical reactions.

Nuclear Chemistry

In This Chapter

- Types of radioactive decay
- Half-lives
- Fusion and fission

Nuclear reactions have many interesting household uses. From the smoke detectors that keep us safe to the electricity that keeps the lights on to the weapons that keep your neighbor's kids out of the backyard, nuclear reactions have a wide variety of uses that help most of us in one way or another.

Unfortunately, not enough people understand nuclear reactions. With this chapter, you, too, will join the ranks of the nuclear-literate.

What Are Nuclear Processes?

Nuclear reactions are reactions that involve the nucleus of an atom. This is very different from regular chemical reactions, because no bonds are created or formed—only the nucleus changes.

We're going to talk about several different forms of chemical reaction, but first, let's review some terms that you're likely to bump into:

- **Nucleons** are the particles in the nucleus of an atom (protons and neutrons).
- **Isotopes** (see Chapter 1) refer to elements with the same number of protons but different numbers of neutrons (resulting in different atomic masses). The different isotopes of an element are referred to as **nuclides.**
- **Radioisotopes** refer to radioactive nuclear isotopes. **Radioactive decay** is when a nucleus gives off various small particles to make it more stable.

Now that we've got the basics, let's start talking about nuclear chemistry!

Chemistrivia

The term **radiation** refers to several different things with very different meanings. **Electromagnetic radiation** refers to wave phenomena such as light or radio waves—for example, every time you turn on your bedroom light you're flooding your room with electromagnetic radiation. **Ionizing radiation** consists of the particles given off during radioactive decay, and is more harmful to health than electromagnetic radiation.

Why Does Radioactive Decay Occur?

Radioactive decay occurs when a nucleus becomes unstable. As a result of this instability, the nucleus needs to change to become more stable. Nobody really understands where this instability comes from, but there are some rules that tend to describe how likely it is that a particular nuclide will be radioactive:

- Atoms with more than 83 protons are radioactive.
- Stable nuclides with low masses tend to have a 1:1 ratio of neutrons to protons, and stable nuclides with higher masses tend to have a ratio that's higher in neutrons. If the ratio of neutrons to protons isn't exactly right, the nuclides tend to be radioactive.
- Nuclides with "magic numbers" of protons or neutrons tend to be more stable than other nuclides. These magic numbers are 2, 8, 20, 50, 82, and 126.
- Nuclides with even numbers of both protons and neutrons are more stable than those with odd numbers of protons and neutrons.

There are exceptions to these rules, which isn't surprising, since nobody really understands why the rules work in the first place. However, if you're not sure if something will be radioactive, these rules are better than random guessing!

Types of Radioactive Decay

There are many ways that atoms can undergo radioactive decay: alpha decay, beta decay, gamma decay, positron emission, and electron capture.

Alpha Decay

Alpha particles (denoted by α) are ordinary helium nuclei ($^{4}_{2}He$) with a charge of +2. When an element undergoes alpha decay, an alpha particle is given off. Alpha emission causes the atomic number of the nuclide undergoing decay to decrease by two and the atomic mass to decrease by four. An example of alpha decay is shown below:

$$^{247}_{97}Bk \rightarrow {}^{4}_{2}He + {}^{243}_{95}Am$$

Chemistrivia

Alpha decay might be responsible for the working of your smoke detector. Many smoke detectors contain a very small amount of americium-241, which gives off alpha particles. When these alpha particles strike the oxygen and nitrogen in the air, they cause the particles to lose electrons and carry a weak electric current. When smoke enters an americium smoke detector, it attaches to these ions and causes the current to stop flowing. Whenever the current in a smoke detector decreases, the detector responds with a piercing shriek to warn of the danger.

Beta Decay

Beta decay (denoted by β) occurs when beta particles (electrons, ${}^{0}_{-1}e$) are emitted from the nucleus of an atom. This process effectively converts a neutron to a proton, increasing the atomic number by one without changing the atomic mass. An example of beta decay is shown below:

$${}^{102}_{39}Y \rightarrow {}^{0}_{-1}e + {}^{102}_{40}Zr$$

Gamma Decay

Gamma decay occurs when very high-energy light (called **gamma rays** and shown by ${}^{0}_{0}\gamma$ in nuclear reactions) is released from the nucleus of an atom. These gamma rays have no mass or charge and are usually given off during the course of other nuclear reactions.

Positron Emission

Positrons are the antiparticles of electrons and have the symbol ${}^{0}_{+1}e$. Like an electron, their mass is essentially zero, but they have a +1 instead of a –1 charge. Positron emission results in the conversion of a proton to a neutron, decreasing the atomic number by one but leaving the atomic mass unchanged:

$${}^{17}_{9}F \rightarrow {}^{17}_{8}O + {}^{0}_{+1}e$$

The Mole Says

Sometimes students get positrons mixed up with protons, because they both have a +1 charge. Protons are made of ordinary matter and have a mass of approximately 1 amu. Positrons are the antimatter equivalent of an electron, so have a mass that's approximately zero.

Electron Capture

Electron capture occurs when an electron in an inner orbital is pulled into the nucleus, converting a proton to a neutron:

$$ {}^{7}_{4}Be + {}^{0}_{-1}e \rightarrow {}^{7}_{3}Li $$

The Mole Says

Like all equations, nuclear equations can be solved using the law of conservation of mass. Using the electron capture example above, you can see that the total mass of the particles to the left of the arrow (7 + 0 = 7 amu) is the same as the mass of lithium—7. Likewise, the charge is also conserved, with the nuclear charge of the left (4 − 1 = +3) equal to the charge in a lithium atom.

Half-Lives

In Chapter 8, we discussed the idea of half-lives. To recap, the half-life of a process is the amount of time it takes for half of the reactant to be converted into products.

When discussing radioactive decay, we also deal with half-lives, except that a nuclear half-life is the amount of time it takes for half of a nuclide to undergo nuclear decay. Because this is basically the same as the kinetics definition, the method used to find the half-life of a radioactive decay process is the same as the method used to find the half-life of a first-order chemical reaction. Our equation, again, is:

$$t_{1/2} = \frac{0.693}{k}$$

where $t_{1/2}$ is the half-life of the process and k is the rate constant for the nuclear decay process.

If you know the half-life of a reaction, you can compute the rate constant, and if you know the rate constant, you can determine the quantity of unreacted nuclide using the first-order rate equation:

$$\ln[A_t] = -kt + \ln[A_o]$$

Let's see a sample problem:

Example: Given that the half-life of ^{236}Pu is 87.74 years, how much of a 175 gram sample will remain after 225 years?

Solution:

1. Determine the rate constant using the first equation:

   $$87.74\,yrs = \frac{0.693}{k}$$

   $k = 7.90 \times 10^{-3}$ / yr

2. Plug the values of k, t, and A_0 into the second equation to solve for A_t.

   $\ln[A_t] = -(7.90 \times 10^{-3} / \text{yr})(225 \text{ yrs}) + \ln(175 \text{ g})$

   $A_t = 29.4$ grams

It's as simple as that!

Chemistrivia

One of the most important uses of half-lives is carbon dating. Living creatures absorb radioactive ^{14}C in the food they eat, incorporating it into their tissues. When they die, the nonradioactive ^{12}C remains, while the radioactive ^{14}C, with a half-life of 5,730 years, slowly vanishes through beta decay. By comparing the quantity of ^{14}C in a sample to the quantity of ^{14}C in living creatures, the ages of formerly living objects can be accurately determined.

Fission and Fusion

While we've so far talked about natural radioactive decay, we science folk have learned how to do some nuclear reactions of our own for the purpose of giving off lots of energy. This energy is used for a number of purposes, including power generation and atomic bombs. Let's take a look at the two types of nuclear reaction we've learned how to use: nuclear fission and nuclear fusion reactions.

Nuclear Fission

Fission occurs when a heavy nucleus is hit with a neutron, causing it to break apart into smaller elements. Because the energy that was used to hold the nucleus together is released during this process, fission gives off a very large quantity of heat. For example, nuclear fission is performed on uranium-235 to generate energy in nuclear power plants:

$$^{1}_{0}n + ^{235}_{92}U \rightarrow ^{137}_{92}Te + ^{97}_{50}Zr + 2\,^{1}_{0}n + heat$$

As you can see, this process generates two neutrons as well as energy and larger nuclides. These two neutrons can, in turn, be used to make other atoms of uranium-235 undergo fission. Because this process is self-perpetuating and repeats itself many times, it's said to be a **chain reaction.** A picture of what this looks like is shown in the following illustration.

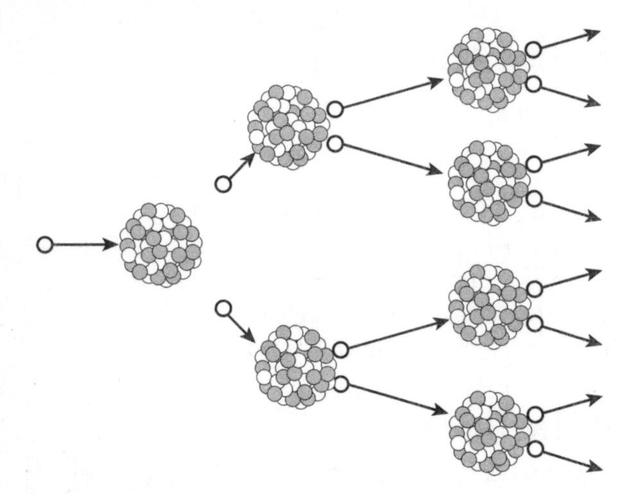

When one uranium atom is broken apart, the neutrons that are formed continue the reaction in neighboring uranium atoms.

Fusion Reactions

Fusion reactions occur when small nuclei stick together to form larger ones. The energy that holds the larger nuclei together is less than the energy needed to hold the smaller nuclei together; this excess energy is given off as heat, as shown below:

$$^{2}_{1}H + ^{3}_{1}H \rightarrow ^{4}_{2}He + heat$$

The only way we really know of to make these reactions occur is to combine the smaller nuclides at temperatures of about 40 million Kelvin. At the current time, the only way we know of to generate this much heat is by fission reactions—as a result, the only practical uses of fusion reactions are in thermonuclear or hydrogen bonds, where a fission reaction is used to initiate a fusion reaction.

Chemistrivia

Because far more energy is released during fusion reactions than fission reactions, a great deal of time and effort has been spent trying to start fusion reactions through less destructive methods than fission bombs. Some of the methods currently being studied include the use of very high-energy lasers or tokamaks (magnetic containment chambers in the generation of high-energy plasmas) to generate the temperatures necessary to initiate a fusion reaction.

Chapter

Electrochemistry

In This Chapter

- Oxidation states
- Redox reactions
- Voltaic cells
- The Nernst equation
- Electrolytic cells

Electrochemistry is responsible for many of the finest things in life. When a child tells his friend to lick a 9-volt battery, electrochemistry is what makes the friend cry. When an office worker talks with his son's principal on the phone about the recent "inappropriate behavior," electrochemistry allows his cubemate to turn up her portable headphones to drown the father out. When a young man gives his sweetheart a ring to celebrate their love, electrochemistry is what disguises the cheap jewelry with a thin layer of gold.

In this chapter, we'll learn more about how to make electrochemistry work for us, too.

Oxidation States

Before we can talk about how electrochemical reactions occur, we need to learn the basics. In electrochemistry, the most basic idea is that of an element's **oxidation state,** which describes the electrical charge that it's considered to have. In simple ionic compounds, the oxidation state is the same as the charge on the ions (for example, in NaCl the oxidation state of Na is +1 and the oxidation state of chlorine is –1).

Things get more complicated when finding the oxidation states of covalent compounds. To find these oxidation states, follow these rules:

1. The oxidation state of a pure element is zero. The atoms in Fe and Cl_2 both have an oxidation state of 0, because only that one element is present in each case.
2. The oxidation state of the most electronegative element in a compound is the same as it would have if it were an anion. For example, in BF_3, the oxidation state of F is –1.
3. The oxidation state of hydrogen is +1, unless it's bonded to a metal, in which case it's –1. For example, in HF it is +1, in NaH it is –1.
4. In neutral compounds, the sum of the oxidation states of all the elements is 0. Going back to the BF_3 example, if each F has an oxidation state of –1, the oxidation state of boron has to be +3 to make the overall charge on the compound 0.

5. In all polyatomic ions, the sum of the oxidation states of all the elements is equal to the charge of the ion. For example, in NH_4^{+1}, the oxidation state of each hydrogen is +1 and the oxidation state of nitrogen is –3, giving an overall +1 charge.

Oxidation and Reduction

During electrochemical processes, the oxidation states of atoms change through the loss or gain of electrons. When an atom loses electrons, the oxidation state becomes more positive, and the atom is said to be *oxidized.* When an atom gains electrons, the oxidation state becomes more negative, and the atom is said to be *reduced.* Reactions in which both oxidation and reduction occur are called *redox reactions.*

Molecular Meanings

Oxidation occurs when an atom loses electrons, and **reduction** occurs when an atom gains electrons. Reactions in which the oxidation states of the elements change are called **redox reactions.**

Let's take a look at a sample redox reaction and see which elements have been oxidized and reduced, using our handy rules for determining oxidation states:

$$2\ Na + ZnCl_2 \rightarrow Zn + 2\ NaCl$$

In this reaction, sodium starts off with an oxidation state of 0 and ends up with an oxidation state of +1. Because the oxidation state is more positive, it has been oxidized. Zinc starts with a +2 oxidation state and ends up with a 0 oxidation state. Because the oxidation state is more negative, it's said to be reduced. Chlorine has a –1 oxidation state through the whole reaction, so it's been neither oxidized nor reduced.

In redox reactions, there must be both an oxidation and a reduction reaction occurring. After all, when an atom gains electrons, those electrons have to come from *somewhere.* Because atoms that gain electrons have pulled them away from atoms that have been oxidized, the atoms that gain electrons are called **oxidizing agents.** Likewise, the elements that lose electrons give them to other elements, which are reduced; we call the atoms that lose electrons **reducing agents.** In our example above, sodium is the reducing agent (it caused zinc to be reduced), and zinc chloride is the oxidizing agent (it caused sodium to be oxidized).

Chemistrivia

If you like lions, an easy way to remember oxidation and reduction is this: "LEO goes GER," which stands for Loss of Electrons = Oxidation, Gain of Electrons = Reduction. Additionally, elements that are oxidized are reducing agents; elements that are reduced are oxidizing agents.

Balancing Redox Equations

Unfortunately, redox reactions are somewhat more complicated to balance than the chemical equations we balanced in Chapter 4. Fortunately, we're going to learn an easy way to balance them, called the **half-reaction method.** Though this method has a lot of steps, it's not too hard once you get the hang of it.

Example: Balance the following redox reaction:

$$Al + MnO_2 \leftrightarrow Mn + Al_2O_3$$

Solution: To balance redox reactions with the half-reaction method, follow these steps:

1. Break the unbalanced equation into two smaller equations called **half-reactions.** The first half-reaction will follow the oxidation of the element that lost electrons, and the second will follow the reduction of the element that gained electrons.

Because aluminum is oxidized in this reaction, the oxidation half-reaction is:

$Al \rightarrow Al_2O_3$

The half-reaction for the reduction of manganese is:

$MnO_2 \rightarrow Mn$

2. In each half-reaction, balance the elements that are oxidized or reduced. After you're done with that, balance any elements other than oxygen or hydrogen.

   For the oxidation half-reaction, we balance the aluminum:

   $2\ Al \rightarrow Al_2O_3$

   For the reduction half-reaction, we don't need to balance anything, because there is only one manganese atom on both the reactant and product side.

3. Balance the oxygen atoms by adding H_2O. We can add water to these equations because most redox reactions take place in water. However, we can still do this for reactions that don't take place in water, because in those cases the water molecules will eventually cancel each other out.

   For the oxidation half-reaction, we do this by adding three water molecules to the reactant side of the equation:

   $2\ Al + 3\ H_2O \rightarrow Al_2O_3$

For the reduction half-reaction, we add two water molecules to the product side of the reaction:

$MnO_2 \rightarrow Mn + 2\ H_2O$

4. Balance the hydrogen atoms for both half-reactions by adding H^+ ions.

   This is done for each half-reaction in the following way:

   $2\ Al + 3\ H_2O \rightarrow Al_2O_3 + 6\ H^+$

   $MnO_2 + 4\ H^+ \rightarrow Mn + 2\ H_2O$

The Mole Says

The reason we can just throw in H^+ ions is because most redox reactions take place in acidic solutions. However, a few redox reactions take place in basic solutions. For these reactions, an additional step is necessary between steps 4 and 5, where hydroxide ions (OH^-) are added to neutralize the phantom "acid" added in step 4. When they neutralize, they form additional water.

5. Add electrons so the total amount of charge on both sides of the reaction is neutral.

Because the equations each have an overall positive charge, we need to add electrons to neutralize the positive ions:

$2\ Al + 3\ H_2O \rightarrow Al_2O_3 + 6\ H^+ + 6\ e^-$

$MnO_2 + 4\ H^+ + 4\ e^- \rightarrow Mn + 2\ H_2O$

6. Multiply the coefficients for both half reactions so the number of electrons in each is the same.

   In our example, there are six electrons in the oxidation half-reaction and four in the reduction half-reaction. To make both numbers of electrons the same, multiply the coefficients in the first half-reaction by two and the second by three:

   $4\ Al + 6\ H_2O \rightarrow 2\ Al_2O_3 + 12\ H^+ + 12\ e^-$

   $3\ MnO_2 + 12\ H^+ + 12\ e^- \rightarrow 3\ Mn + 6\ H_2O$

7. Add the two half-reactions together.

   $4\ Al + 6\ H_2O + 3\ MnO_2 + 12\ H^+ + 12\ e^- \rightarrow$
   $2\ Al_2O_3 + 12\ H^+ + 12\ e^- + 3\ Mn + 6\ H_2O$

8. Cancel out any terms that are present on both sides of the resulting equation, to give us our answer:

   $4\ Al + 3\ MnO_2 \leftrightarrow 2\ Al_2O_3 + 3\ Mn$

Voltaic Cells

Now that we know about electrochemical reactions, it's time to start talking about electrochemistry you may be familiar with—batteries. As it turns out, a fancy word for "battery" is **voltaic cell**, an example of which is shown next.

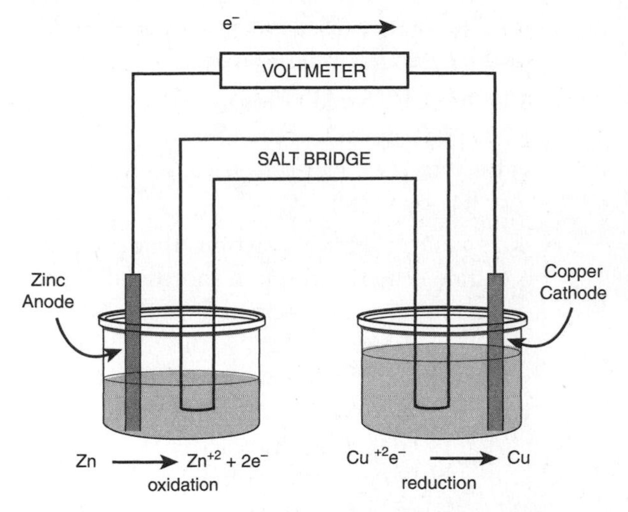

A voltaic cell in which zinc is oxidized and copper is reduced.

Before we can really understand how voltaic cells work, we need to understand some of the basic terminology:

- The pieces of metal that are dipped into the solutions in this diagram are called **electrodes.** The electrode at which oxidation takes place is called the **anode** (in our example, zinc), and the electrode at which reduction occurs is called the **cathode** (in our example, Cu).
- The U-shaped tube between the two beakers in this diagram is called a **salt bridge.** Salt bridges contain an electrolyte solution that conducts electricity and won't react with any of the other chemicals in the cell. Salt bridges are required because the charge carriers in a voltaic cell are ions, and

for electricity to be conducted, the two sides of the cell need to be connected.

- The term **half-cell** refers to the process that takes place at each of the electrodes. Every battery has two half-cells because every battery has two electrodes.
- The electrochemical reactions that occur in a voltaic cell are written in shorthand. The oxidation reaction is written first—in our cell, zinc is converted to Zn^{+2}, so the reaction is written as $Zn | Zn^{+2}$. The reduction reaction is written second—in our cell, Cu^{+2} is converted to copper, so the reaction is written $Cu^{+2} | Cu$. To write the overall process for the cell, write these reactions in order, with a double vertical line separating them: $Zn | Zn^{+2} || Cu^{+2} | Cu$.

Standard Electrode Potentials

When engineers design a battery, they can't just put any two solutions into the cell and expect it to work correctly. One of the most important factors when designing a battery is the **voltage.** Voltage is a measure of how forcefully electrons are moved from one place to another, and is defined as the amount of energy given off by a spontaneous electrochemical process (or the amount of energy needed to make a nonspontaneous electrochemical process occur).

In order to calculate the **standard cell potential,** you would use the following equation:

$$E^{o}_{cell} = E^{o}_{oxidation} + E^{o}_{reduction}$$

This means that to find the potential of a cell, you need to add the potentials of the reactions that take place at the anode and cathode half-cells. (Incidentally, the little ^{o} above each E value indicates that the cell is running with a solution concentration of 1 M. This won't always be the case, but we'll worry about that in the next section, "The Nernst Equation.")

How do we figure out these half-cell potentials? We don't. Instead, we look them up in a table, like the one shown below:

Standard reduction potential (V)	Reduction half-reaction
2.87	$F_{2(g)} + 2\ e^{-} \rightarrow 2\ F^{-}_{(aq)}$
0.34	$Cu^{+2}_{(aq)} + 2\ e^{-} \rightarrow Cu_{(s)}$
0.00	$2\ H^{+}_{(aq)} + 2\ e^{-} \rightarrow H_{2(g)}$
–0.44	$Fe^{+2}_{(aq)} + 2\ e^{-} \rightarrow Fe_{(s)}$
–0.76	$Zn^{+2}_{(aq)} + 2\ e^{-} \rightarrow Zn_{(s)}$

You may have noticed that this chart lists only reduction potentials, not oxidation potentials. Fortunately, you can find the oxidation potential of a half-reaction by reversing the sign of the reduction potential. For example, by reversing the first entry in this chart, we find that the standard oxidation potential for the reaction $2\ F^{-} \rightarrow F_{2} + 2\ e^{-}$ is –2.87 V.

Now that we know how to find the potential of a voltaic cell, let's do it for the cell we saw earlier: $Zn | Zn^{+2} || Cu^{+2} | Cu$.

The following two half-reactions take place:

$Zn \rightarrow Zn^{+2} + 2\ e^-$ oxidation

$Cu^{+2} + 2\ e^- \rightarrow Cu$ reduction

To find the overall cell potential, we add the half-cell potentials. For the oxidation of zinc to Zn^{+2}, the half-cell potential is exactly the same as for the reduction of zinc, except with the opposite sign—0.76 V. For the reduction of Cu^{+2} to pure copper, it's 0.34 V. When you add them up using the equation for standard cell potential, you find that:

$$E^{o}_{cell} = E^{o}_{oxidation} + E^{o}_{reduction}$$
$$= 0.76\ V + 0.34\ V$$
$$= 1.10\ V$$

Chemistrivia

You may have noticed that there is no combination of half-reactions that will result in a standard cell potential of 9 V. In the common 9 V batteries you have around your house, there are actually several batteries with smaller voltages that are connected to give a total voltage of 9 V.

The Nernst Equation

The preceding calculations are handy when all of the reactions take place under standard conditions (1 M for solutions, 1 atm for gases). However, when this isn't possible, we bring in **the Nernst equation** to find the voltage:

$$E = E^{\circ} - \frac{0.0591}{n} \log Q$$

In this equation, E is the cell potential under the conditions given, E° is the standard cell potential, n is the number of electrons transferred in the reaction, and Q is the reaction quotient.

The Mole Says

The reaction quotient for the generic process aA + bB → cC + dD, is:

$$Q = \frac{[C_o]^c[D_o]^d}{[A_o]^a[B_o]^b}$$

Where C_o is the initial molarity of C, D_o is the initial molarity of D, and so on. For gaseous equilibria, partial pressures (in atm) should be used in lieu of molarities.

Example: For the cell $Zn|Zn^{+2}||Cu^{+2}|Cu$, what is the cell potential if the concentration of Zn^{+2} is 2.5 M and the concentration of Cu^{+2} is 0.75 M?

Solution: In the last section, we found that E°_{cell} for this process was 1.10 V. However, before using the Nernst equation, it is necessary to figure out what values we should use for all the variables.

For this process, n = 2 because two electrons are transferred from Zn to Cu^{+2} whenever this reaction takes place.

To find the reaction quotient, we need to write out the equation for the entire process:

$$Zn_{(s)} + Cu^{+2}_{(aq)} \rightarrow Zn^{+2}_{(aq)} + Cu_{(s)}$$

$$Q = \frac{[Zn^{+2}][Cu]}{[Zn][Cu^{+2}]}$$

In Chapter 9, we learned that we don't need to include pure solids in equilibrium expressions, so we'll leave out the neutral copper and zinc here. As a result, Q is:

$$Q = \frac{[Zn^{+2}]}{[Cu^{+2}]} = \frac{2.5M}{0.75M} = 3.3$$

Putting all of these values into the Nernst equation, we find that:

$$E = 1.10V - \frac{0.0591}{2}\log 3.3$$

$$E = 1.10V - 0.015\ \text{V}$$

$$E = 1.08V$$

Electrolytic Cells

If you're familiar with the home shopping channels, you know electroplating is a big business. For those of you with better ways to spend your time, electroplating is a process by which a very thin coating of a precious metal is placed over a much cheaper metal, as in the case of cheap jewelry.

Electroplating is made possible by electrolytic cells. Electrolysis is a process by which a current is forced through a cell to make a nonspontaneous electrochemical change occur. For example, by forcing electricity through a cell, we can force electrochemical reactions with negative cell potentials to occur. An electrolytic cell is shown below:

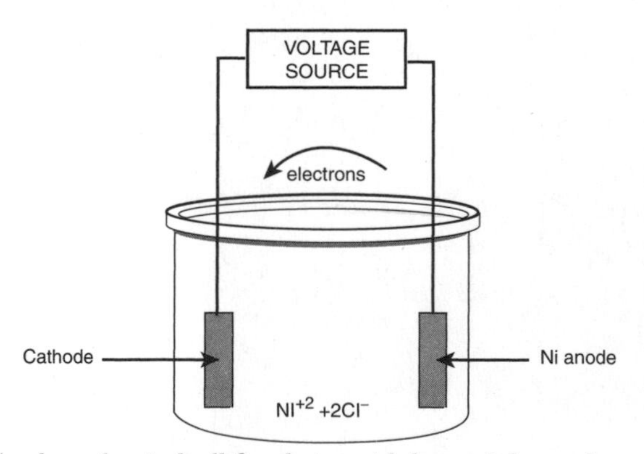

An electrochemical cell for plating nickel on stainless steel.

In this electrochemical cell, the following takes place:

1. $NiCl_2$ dissolves to form Ni^{+2} and Cl^{-1} ions.
2. When the cell is turned on, Ni^{+2} ions move toward the cathode, and Cl^{-1} ions move toward the anode.

3. At the cathode, the Ni^{+2} ions are reduced to form a thin layer of nickel. Typically, the item to be electroplated (for example, cheap jewelry) is used as the cathode.
4. During this process, the nickel anode will oxidize, generating more Ni^{+2} ions to replace those that were plated onto the cathode.

During this process, the anode eventually disappears, having been electroplated onto the cathode.

Chapter 

Thermodynamics

In This Chapter

- Energy changes
- Enthalpy
- Entropy and randomness
- Free energy

One thing we haven't really discussed much is energy. We know that energy is given off if you set your neighbor's car on fire, but we haven't really discussed this from a chemistry standpoint. Now, that's all about to change!

Some Basic Thermodynamic Terms

Thermodynamics is the study of heat. However, before we can understand what this is all about, we need to first define some important terms that we'll need for the rest of this chapter:

- **Energy** is the capacity of something to do work or produce heat. Energy can be broken down into two forms: potential energy and kinetic energy. **Potential energy** is stored energy; before my nephew has a chocolate bar, a great deal of potential energy is stored within his delicious snack. **Kinetic energy** is energy having to do with motion; after my nephew eats the chocolate, the resulting mayhem he unleashes on the world represents kinetic energy.

The Mole Says

The metric unit for energy is the joule (J). Another standard unit for energy is the calorie. Calories (with a capital C) are used to measure the energy in food; 1 Calorie (capital C) = 1,000 calories (lowercase C).

- **The law of conservation of energy,** also known as **the first law of thermodynamics,** states that energy is never created nor destroyed in any process. This means that while we can convert potential energy to kinetic energy (or vice-versa), the overall amount of energy is the same.
- **Temperature** describes the amount of motion that the molecules or atoms in a material have. Fast movement represents high temperature, and slow movement represents low temperature.

- **Heat** describes the amount of energy that is transferred from one object to another. Heat depends on temperature, because if two objects have different temperatures, the very fast motion of the particles in one material will eventually be transferred into the second material.

Bad Reactions

Remember, heat and temperature are not the same thing! Though heat is highly dependent on temperature, temperature is a measure of motion within a material, and heat is a measure of energy change between two materials.

- The symbol Δ is often used in thermodynamics to denote change. For example, if the energy of something increases, we'd say that "ΔE is positive."
- The symbol ° above a term indicates that the term is valid under standard conditions of 1 atm and 298 K (25° C). For example, the term ΔH°_f refers to a "standard heat of formation," about which we'll learn more later in this chapter.
- The **system** in thermodynamics refers to whatever is being studied, and the **surroundings** are everything else in the universe. If we place a can of beans into a

campfire, the beans are the system and the surroundings are everything else in the entire universe (though we'd probably only consider the campfire, for simplicity's sake).

Now that we've gotten that out of the way, let's learn some thermodynamics!

Energy Change

All materials contain some quantity of energy, which we'll represent by E. As we mentioned before, this quantity of energy can change. For example, if I throw a can of beans into a campfire, the energy of the system increases (ΔE is positive). If we place a can of beans into a pot of ice water, the energy of the system decreases (ΔE is negative). For all processes, the change in energy, ΔE, is equal to the difference in the energy of the system before the process and the energy of the system after the process. In equation form, this is:

$$\Delta E = \Delta E_{final} - \Delta E_{initial}$$

If you'll recall, energy is defined as the capacity of a system to produce heat or do work. As a result, we can say that the energy change (ΔE) for a process is equal to the change in heat for the process (q) and the amount of work performed by/on the system (w):

$$\Delta E = q + w$$

If the change in heat for a process is positive, it is said to be **endothermic** because the process won't occur without the addition of energy. If the change in heat is negative, the process is **exothermic** because the process causes the system to release heat into its surroundings.

The Mole Says

Generally, processes that feel cold (like an ice pack) are endothermic and processes that feel hot (like a fire) are exothermic.

Enthalpy

One type of work common in chemical processes has to do with the expansion or contraction of gases. In the example of an automobile, the expansion of gases in the cylinders causes a piston to move, which ultimately causes the car to move forward. This sort of work is shown below:

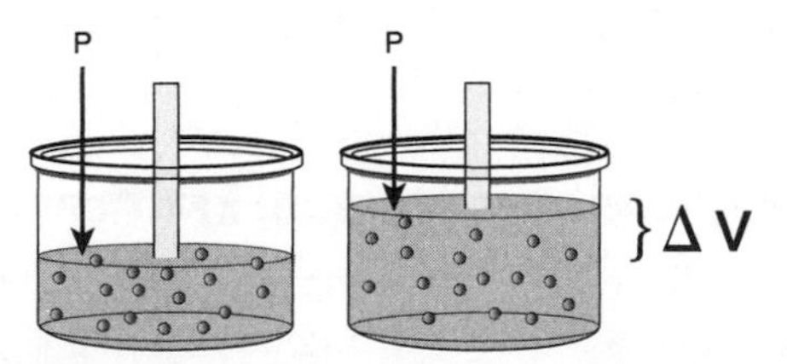

When the gas in a cylinder expands, the product of the change in volume (ΔV) and the outside pressure (P) is equal to the amount of work performed.

When a gas in a cylinder performs work, the amount of work is equal to the product of the change in volume and the outside pressure (assuming the outside pressure is constant). In equation form, this is:

$$w = -P\Delta V$$

In this expression, the sign is negative because the gas is performing work on the system, causing it to lose energy. If we substitute this term into the equation for ΔE we saw earlier, we get the expression:

$$q_p = \Delta E + P\Delta V$$

The small p under the q indicates that this is the change in heat when the pressure remains constant outside the cylinder. Because this change in heat is such an important thing in chemistry, we give it a special term, called **enthalpy,** which is represented by the term ΔH. Substituting this for q_p, we get the equation:

$$\Delta H = \Delta E + P\Delta V$$

Because enthalpy is such a special term, there are many different uses for it in chemical calculations, among them the following.

Enthalpy Changes in Chemical Reactions

The enthalpy change for any process is equal to the difference between the starting and ending enthalpies of all of the things taking part in the process. In equation form, this is shown as:

$$\Delta H = H_{final} - H_{initial}$$

Typically, a symbol is written under ΔH to indicate what type of process is taking place. Two of the most common are:

- ΔH_{rxn} indicates the change in enthalpy for a chemical reaction. The equation in this case is

  $\Delta H_{rxn} = \Delta H_{products} - \Delta H_{reactants}$.
- ΔH_f indicates the amount of energy needed to form a substance from its constituent elements.

Any other subscript you see after a ΔH term gives you information about the process taking place. Don't worry, they should be easy to figure out from the information you'll be given.

To understand what this all means, let's do a sample problem:

Example: Find the standard heat of combustion of ethene given that $\Delta H^\circ_f(C_2H_{4(g)}) = +52.3$ kJ/mol, $\Delta H^\circ_f(CO_2) = -393.5$ kJ/mol, $\Delta H^\circ_f\,(H_2O_{(l)}) = -285$ kJ/mol, and the equation:

$$C_2H_{4(g)} + 3\ O_{2(g)} \rightarrow 2\ CO_{2(g)} + 2\ H_2O_{(l)}$$

Solution: We need to solve the following equation:

$$\Delta H^\circ_{rxn} = H_{products} - H_{reactants}$$

The total heat of formation of the products is equal to twice the heat of formation of CO_2 (2 × –393.5 kJ/mol = –787 kJ/mol) plus twice the heat of formation of water (2 × –285.3 kJ/mol = –571.6 kJ/mol) for a total of –1358.6 kJ/mol.

The total heat of formation of the reactants is equal to the heat of formation of ethene (+52.3 kJ/mol) plus three times the heat of formation of oxygen (0 kJ/mol, since it's in its most stable form) for a total of +52.3 kJ/mol.

Subtracting the heat of formation for the reactants from the heat of formation for the products, we find that the standard heat of combustion for ethene is –1410.9 kJ/mol.

Hess's Law

Sometimes the reactants in a chemical reaction need to undergo several steps before forming products. For such processes, the sum of the enthalpy changes for each step is equal to the overall enthalpy for the process. This is known as **Hess's Law.**

To see how this works, let's find the overall energy change and equation for the process A + B + 2 D → E, given the following equations:

$$A + B \rightarrow C \qquad \Delta H^\circ_{rxn} = -100 \text{ kJ/mol}$$
$$E \rightarrow C + 2\,D \qquad \Delta H^\circ_{rxn} = +300 \text{ kJ/mol}$$

To figure this out, all we need to do is add these equations together in such a way that we get the equation we're looking for. Because A, B, and D are on the left and E is on the right, we'll add these equations together in a way that gives us this result:

$A + B \rightarrow C \quad \Delta H^\circ_{rxn} = -100$ kJ/mol

$C + 2\ D \rightarrow E \quad \Delta H^\circ_{rxn} = -300$ kJ/mol

---

$A + B + C + 2\ D \rightarrow C + E \quad \Delta H^\circ_f = (-100 + -300)$kJ/mol

The Mole Says

You'll notice that when we reversed the second reaction, the ΔH°_{rxn} term changed signs, from positive to negative. This indicates that if the reaction goes from $E \rightarrow C + 2\ D$ the reaction requires energy to occur (is endothermic), while the reverse process $C + 2\ D \rightarrow E$ releases energy (is exothermic). Whenever you reverse an equation like this, make sure you also change the sign on ΔH°_{rxn}!

When we simplify this by removing the C term (because it's on both sides of the equation, causing it to cancel out), we find that:

$$A + B + 2\ D \rightarrow E \quad \Delta H^\circ_f = -400 \text{ kJ/mol}$$

Calorimetry

Calorimetry is the process by which the energy change of a chemical reaction can be experimentally determined. It works by performing a chemical reaction in a giant steel container called a "bomb" that's immersed in a giant bucket of water. The energy that is generated by the reaction in the bomb is transferred to the water, the temperature of which can be easily measured. A bomb calorimeter is shown below:

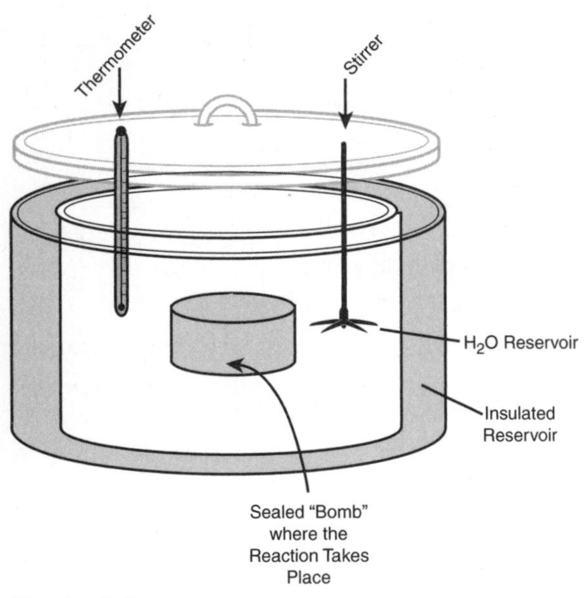

A bomb calorimeter.

The heat released by a reaction within a bomb calorimeter can be calculated using the equation:

$$\Delta H_{rxn} = mC_p\Delta T$$

where ΔH_{rxn} is the amount of energy transferred from the bomb to the water, m is the mass of the water (in grams), C_p is the heat capacity of water (4.184 J/g°C), and ΔT is the change in water temperature (in °C). Most commonly, bomb calorimetry is used to find the heat of combustion ($\Delta H_{combustion}$) for a process.

For example, if a reaction caused the temperature of 525 grams of water to increase by 12.5° C, the heat released by the reaction would be:

$$\Delta H_{rxn} = (525.0\text{ g})(4.184\text{ J/g°C})(12.50°\text{ C}) = 27{,}460\text{ J}$$

It's a simple as that!

Entropy

You may have noticed that some events don't happen very often. For example, it's uncommon to see apples fall upward and attach to trees. The reason for this is that these processes aren't spontaneous. Other things, such as the trickling of water down your roof after a rainstorm, are spontaneous.

Why do spontaneous events occur? Spontaneous events take place because they make the world more random. For example, imagine opening a bottle of toxic gas in your living room. Do you believe that the gas will stay in a corner of your room instead of filling the room? Of course not! The gas will fill the room because by spreading over a larger area, the randomness of the gas molecules is larger than if they stay in the same place. The randomness of a system is called **entropy,** denoted by the symbol S.

This behavior is spelled out in the **second law of thermodynamics,** which states that for any process, the entropy of the universe increases. This may seem counterintuitive; after all, if you put away your laundry, haven't you made the universe more orderly? Actually, you haven't, because the randomness that you eliminated by putting away your laundry was more than made up for by extra randomness caused by your body breaking down food to give off the energy needed to do it. It's possible to decrease the randomness of a system, but overall, the universe as a whole will have more randomness than when you started.

The Mole Says

Of the phases of matter, gases (which are highly disordered) have more entropy than liquids (which have some disorder), which have more entropy than solids (which are highly ordered).

To calculate the change in entropy for a process (ΔS), you do exactly the same calculation that you used to calculate enthalpy changes (except, of course, that the equation contains S and not H):

$$\Delta S = S_{final} - S_{initial}$$

Also, as with enthalpy changes, this equation frequently takes the following form for chemical equations:

$$\Delta S^{o}_{rxn} = S^{o}_{products} - S^{o}_{reactants}$$

Free Energy

As we saw in the preceding section, spontaneous processes take place if they cause the entropy of the universe to increase. This is simple enough to understand, but not so simple to calculate, because we rarely know much about the entire universe.

As a result, there's another term that we use to determine whether a chemical process will be spontaneous by itself, without help from the rest of the universe. This term is called **free energy,** and is denoted by the term G. Free energy is calculated assuming that the pressure and temperature of the system being studied are constant, a good assumption for many chemical reactions.

Chemistrivia

The G in free energy refers to Josiah Gibbs, for whom this term is frequently referred to as "Gibbs free energy."

If the change in free energy (ΔG) is negative, the process will be spontaneous. If ΔG is positive, the process is nonspontaneous. If ΔG is zero, the system is at equilibrium.

Relating Free Energy to Entropy and Enthalpy

The change in free energy can be related to the entropy and enthalpy changes for a process using the equation:

$$\Delta G = \Delta H - T\Delta S$$

The ΔH term in this equation reflects the fact that exothermic processes are more likely to be spontaneous than endothermic processes. The ΔS term reflects the fact that a process with positive entropy tends to be spontaneous.

As we can see from the chart below, the temperature term in this equation plays a large part in determining whether a process is spontaneous or not:

ΔH	ΔS	ΔG	Is it spontaneous?
+	+	+ at low temperatures – at high temperatures	no, at low temperatures yes, at high temperatures
+	–	+ at all temperatures	no, at all temperatures
–	+	– at all temperatures	yes, at all temperatures
–	–	– at low temperatures + at high temperatures	yes, at low temperatures no, at high temperatures

If you're given ΔH, ΔS, and T, it's a fairly simple matter to determine whether a process is spontaneous using the equation $\Delta G = \Delta H - T\Delta S$.

Example: Find the change in free energy for the following reaction at 298 K:

$$2\ H_{2(g)} + O_{2(g)} \rightarrow 2\ H_2O_{(g)}$$

given that $\Delta H^\circ_{rxn} = -483.6$ kJ and $\Delta S^\circ_{rxn} = -88.8$ J/K.

Solution: To solve this problem, all you need to do is plug the values you've been given into the equation for ΔG. In this case:

$$\Delta G^\circ_{rxn} = -483.6\ \text{kJ} - (298\ \text{K})(-88.8\ \text{J/K})$$
$$= -483.6\ \text{kJ} - (-26.5\ \text{kJ})$$
$$= -457.1\ \text{kJ}$$

Using this information, we can see that this reaction will be spontaneous at 298 K.

Bad Reactions

The reaction in the example above actually takes place too slowly to measure at 298 K, despite the fact that it's a spontaneous reaction. When doing calculations for ΔG_{rxn}, keep in mind that while they tell you whether a reaction is spontaneous, they don't say anything at all about how quickly the reaction will occur. For that, you need kinetic data (see Chapter 8).

To calculate the change in free energy for a process (ΔG), you perform exactly the same calculation that you used to calculate enthalpy and entropy changes (except that the equation contains G):

$$\Delta G = G_{final} - G_{initial}$$

As with enthalpy and entropy changes, this equation takes the following form for chemical reactions:

$$\Delta G^o_{rxn} = G^o_{products} - G^o_{reactants}$$

Relating Free Energy to Equilibrium Constants

By looking at the magnitude of ΔG^o for a process, we can get a pretty good feel for what the equilibrium constant, K (see Chapter 9), is for the reaction.

- When $\Delta G^o = 0$, the reaction is at equilibrium. As a result, the equilibrium constant K is 1.
- When ΔG^o is negative, the reaction is spontaneous in forming products from reactants, so the equilibrium constant K is greater than 1.
- When ΔG^o is positive, the reaction is nonspontaneous in forming products from reactants but spontaneous in the reverse direction. As a result, the equilibrium constant K is less than 1.

From this information, we can tell a great deal about the free energy of a process from an equilibrium constant. For example, if the equilibrium constant is 3,472, the free energy for the reaction will be very negative, indicating a spontaneous process. Likewise, if the equilibrium constant is 0.0000028, the free energy for the process will be very positive, indicating a nonspontaneous process.

Glossary

acid Any material that can accept a pair of electrons. In aqueous solutions, those with pH < 7.00.

acid-base reaction An electron pair is donated by a base to an acceptor acid. In aqueous solution, the reaction of H^+ and OH^- yields water.

acid dissociation constant (K_a) The constant that describes the equilibrium position for the dissolution of an acid in water.

activation energy The minimum amount of energy that's required for the reactants of a chemical reaction to form products.

alkane A hydrocarbon that contains only single carbon-carbon bonds. Also called a "saturated hydrocarbon."

alkene A hydrocarbon that contains at least one double bond.

alkyne A hydrocarbon that contains at least one triple bond.

alloy A metal in which several elements are present.

alpha decay When a nucleus breaks apart, emitting a helium nucleus, which is called an "alpha particle" in this context.

amorphous solid A solid material in which the molecules have no long-range order.

amu Atomic mass unit, equivalent to $\sim 1.67 \times 10^{-27}$ kg.

angular momentum quantum number Denoted by *l*, it determines the shape and type of the orbital. Possible values are 0, 1, 2, … (n–1).

anion An atom or group of atoms with negative electrical charge.

aromatic hydrocarbon A hydrocarbon containing alternating single C-C and double C-C bonds in a ring. The best known aromatic hydrocarbon is benzene (C_6H_6).

Arrhenius theory of acids and bases A compound that forms hydronium (H^+) ions in water is an acid, and a compound that forms the hydroxide ion in water is a base.

atom The smallest chunk of an element with the same properties as larger chunks of that element.

atomic mass The sum of the number of protons and number of neutrons in the nucleus of an atom, denoted by the symbol A.

atomic number The number of protons in an element, denoted by the symbol Z.

atomic radius One-half the distance between the nuclei of two bonded atoms of the same element.

Aufbau principle Electrons fill orbitals with lower energy before they fill those with higher energies.

Avogadro's number 6.02×10^{23}.

base Any molecule that can donate a pair of electrons to form a bond. In aqueous solutions, those with $pH > 7.00$.

beta decay When an electron (called a "beta particle" in this context) is emitted during the radioactive decay of an atomic nucleus.

Brønsted-Lowry theory of acids and bases A compound that gives H^+ ions to other compounds is an acid, and a compound that accepts H^+ ions from other compounds is a base.

buffer A solution consisting of a weak acid and its conjugate base that resists changes in pH when acid or base is added to it.

catalyst A material that increases the rate of a chemical reaction without being consumed.

cation An atom or group of atoms with a positive electrical charge.

cell potential A measure of the electromotive force that drives electrons in a voltaic cell.

colligative property Any property of a solution that depends on the concentration.

colloid Stable materials in which one type of particle is suspended in another without actually having been dissolved.

combustion When organic molecules combine with oxygen to form carbon dioxide, water vapor, and a large quantity of heat.

condensation The process by which a gas becomes a liquid.

conductor A material through which electricity can flow.

conjugate acid-base pairs A conjugate acid is the compound formed when a Brønsted-Lowry base accepts a proton, and a conjugate base is the compound formed when a Brønsted-Lowry acid gives up a proton.

covalent bond Bonds created when two valence electrons are shared.

covalent compound Compound created when two or more atoms are held together with covalent bonds.

critical point The conditions of pressure ("critical pressure") and temperature ("critical temperature") past which the gas and liquid phases of a material can no longer be distinguished from one another.

crystal Large arrangements of ions or atoms that are stacked in regular patterns.

decomposition reaction When large molecules break apart to form smaller molecules.

deposition The process by which a gas becomes a solid without first becoming a liquid.

differential rate law A rate law that explains the relationship between the concentration of the reactants and the reaction rate.

dilution The process by which a solvent is added to a solution to make the solution less concentrated.

dipole-dipole force An attractive force caused when the partially negative side of one polar molecule interacts with the partially positive side of another.

double displacement reaction A reaction that occurs when the cations of two ionic compounds switch places.

electrode The location of oxidation or reduction in a voltaic cell. The cathode is the electrode at which reduction occurs, and the anode is the electrode at which oxidation occurs.

electrolysis The process by which a current is forced through a cell in order to make a nonspontaneous electrochemical change occur.

electrolyte A compound that, when dissolved, causes water to conduct electricity.

electron Negatively charged particles that are found in the orbitals outside the nucleus of an atom.

electron affinity The energy change that occurs when a gaseous atom picks up an extra electron.

electron capture When an inner shell electron is captured by the nucleus, decreasing the atomic number by one.

electron configuration A list of orbitals that contain the electrons in an atom.

electronegativity A measure of how much an atom will tend to pull electrons away from other atoms to which it has bonded.

endpoint The point where you stop a titration, generally because an indicator has changed color.

endothermic A reaction that requires energy to occur.

energy The capacity of an object to do work or produce heat.

energy diagram A graph that shows the amount of energy that the reactants have at all points throughout the chemical reaction.

enthalpy (H) The amount of heat present in a system at constant pressure.

entropy (S) A measure of the randomness of a system.

equilibrium When the concentrations of the products and reactants of a chemical reaction have stabilized because the rates of the forward and backward processes are the same.

equilibrium constant (K_{eq}) A constant that indicates whether the equilibrium will lie toward products or reactants.

equivalence point The point in a titration where $[H^+] = [OH^-]$.

excess reactant In a limiting reactant problem, the reactant that is left over when the limiting reactant has been completely used up.

excited state Any orbital with higher energy than the ground state.

exothermic A reaction that releases heat.

fission When an atomic nucleus breaks apart to make two smaller ones and a huge amount of energy.

free energy (G) Gibbs free energy, which is composed of enthalpy (heat) and entropy (randomness). G is the fundamental measure that determines the position of equilibria and the rates of reactions. It is usually expressed in kJ/mol.

fusion A nuclear process in which small nuclei combine to make larger ones plus a huge quantity of energy.

gamma ray Very high-energy electromagnetic radiation that's frequently given off when a nucleus undergoes radioactive decay.

ground state The orbital in which an electron is found if energy is not added to the atom.

group A column in the periodic table. Elements in the same group have similar chemical and physical properties.

half-cell The chemical process that takes place at one of the electrodes in a voltaic cell.

half-life ($t_{1/2}$) The amount of time it takes for half of the reactant to be converted to product in a first-order chemical or nuclear process.

half-reaction A reaction that shows only the oxidation or reduction process in a redox reaction.

heat The amount of energy that is transferred from one object to another during some process.

homogeneous mixture A mixture created when two or more substances are so completely mixed with one another that it has uniform composition. Also known as a solution.

Hund's rule Electrons will stay unpaired whenever possible in orbitals with equal energies.

hybrid orbital An orbital formed by mixing two or more of the outermost orbitals in an atom together.

hydrocarbon A molecule that contains only carbon and hydrogen.

hydrogen bond The attraction between a hydrogen atom that's bonded to nitrogen, oxygen, or fluorine and the lone pair electrons on the nitrogen, oxygen, or fluorine atom of a neighboring molecule.

ideal gas A gas that follows all of the postulates of the kinetic molecular theory.

indicator A compound used to indicate whether a solution is acidic or basic.

insulator A material through which electricity can't flow.

integrated rate law A rate law that describes how the concentrations of the reactants in a chemical reaction vary over time.

intermediate A chemical that was formed by one step in a reaction mechanism that will be consumed in another.

intermolecular force A force that holds covalent molecules to one another.

ion Particle with either positive or negative charge. Anions have negative charge and cations have positive charge.

ion product constant (K_w) Equal to 10^{-14}, it's the product of the H^+ and OH^- concentrations in an aqueous solution.

ionic compound A compound formed when a cation and anion combine with one another.

ionization energy The amount of energy required to pull one electron off of an atom.

isomers Molecules with the same molecular formula but different structural formulas.

isotopes Atoms of the same element that have different masses. These different masses are due to differing numbers of neutrons in the nucleus.

K_a The acid dissociation constant, which describes the position of the equilibrium $HA \leftrightarrow H^+ + A^-$.

kinetic energy Energy caused by the motion of an object.

kinetics The study of reaction rates.

Le Châtelier's Principle If you change the conditions of an equilibrium, the equilibrium will shift in a way that minimizes the effects of whatever it is you did.

Lewis acid/base theory A compound that can accept electron pairs from another compound is a Lewis acid, and a compound that can donate electron pairs to another compound is a Lewis base.

Lewis structure A picture that shows all of the valence electrons and atoms in a covalently bonded molecule.

limiting reactant The reactant that runs out first in a chemical reaction, limiting the amount of product that can be formed.

London dispersion forces Temporary dipole-dipole forces created when one molecule with a temporary dipole induces another to become temporarily polar.

magnetic quantum number Denoted by m_l, it determines the direction that the orbital points in space. Possible values for m_l are all the integers from $-l$ through l.

mechanism The process through which reactants form products.

molality (m) Moles of solute per kilograms of solvent.

molar mass The weight of 6.02×10^{23} atoms or molecules of a compound in grams.

molar volume The volume of one mole of any gas at standard temperature and pressure.

molarity (M) Moles of solute per liters of solution.

mole 6.02×10^{23} things.

mole fraction (χ) The number of moles of one component in a solution divided by the total number of moles of all components in the mixture.

mole ratio The ratio of moles of product to the ratio of moles of reactant of a chemical reaction.

molecular solid A material consisting of many covalent molecules held together by intermolecular forces.

molecule A group of atoms held together with covalent bonds.

network atomic solid A material in which many atoms are bonded together covalently to form one gigantic molecule.

neutrons Neutral particles with a mass of about 1 amu that are found in the nucleus of an atom.

normality (N) Number of moles of a reactive species per liter of solution.

nucleon The particles in the nucleus of an atom, namely protons and neutrons.

nucleus The center of an atom where the protons and neutrons are found.

octet rule Elements tend to want to gain or lose electrons to attain the same electron configurations as the nearest noble gas.

orbital Regions of space outside the nucleus of an atom in which electrons can be found.

order An exponential term in a rate law that describes how the overall rate of the reaction depends on the concentration of each reactant.

organic compound Covalent compound that contains carbon and hydrogen.

oxidation state Also called the "oxidation number," this is the charge that the atom is considered to have in a chemical compound.

oxidize To lose electrons.

oxidizing agent A compound that causes another to be oxidized. In the process, it is itself reduced.

partial pressure The partial pressure of one gas in a mixture of gases is equal to the amount of pressure that would be exerted by that gas alone if all of the other gases were removed.

Pauli exclusion principle No two electrons in an atom can have the same four quantum numbers.

period A horizontal row in the periodic table. Elements in the same period have valence electrons with similar energies.

periodic trend Any property which changes in elements as you move across a period or down a group. Examples include ionization energy, atomic radius, electronegativity, and electron affinity.

pH The scale used to indicate the acidity of a solution, defined as $-\log[H^+]$.

phase diagram A graph that shows in what phase a material can be found at all combinations of temperature and pressure.

pK_a $-\log(K_a)$.

polar A term referring to a molecule that has partial positive charge on one side and partial negative charge on the other.

polar covalent bond A covalent bond in which the electrons aren't shared equally between both atoms.

polyatomic ion An ion containing more than one atom.

potential energy Stored energy. In chemical processes, it's frequently stored in chemical bonds.

principal quantum number Denoted by *n*, it describes the energy level of an electron. Possible values are 1, 2, 3 … n.

protons Positively charged particles with a mass of about 1 amu that are found in the nucleus of an atom.

radiation The small particles emitted during the radioactive decay of an atomic nucleus.

rate constant (k) A constant, unique for every chemical reaction, that indicates how quickly it will form products from reactants.

rate-determining step The elementary step in a reaction mechanism that proceeds most slowly.

rate law An expression that shows how the rate of a chemical reaction depends on the concentration or temperature of the reactants.

reaction order The sums of the orders of all reactants in a chemical reaction.

redox reaction A reaction in which the oxidation state of the reactants changes.

reducing agent A compound that causes another to be reduced. In the process, it is itself oxidized.

reduction The process of gaining electrons.

resonance structures Lewis structures in which the positions of the electrons or bonds in a molecule are changed, but the atoms remain in the same locations.

reversible reaction A reaction in which the reactants form products and the products re-form reactants.

root-mean-square (RMS) velocity The average velocity of the molecules in a gas.

salt bridge A tube containing an ionic compound that allows charge transfer in a voltaic cell.

saturated solution A solution that has dissolved the maximum possible amount of solute.

semiconductor A material through which electricity flows well only at high temperatures or voltages.

single displacement reaction When a pure element switches places with one of the elements in a chemical compound.

solubility product constant (K_{sp}) The equilibrium constant for the dissociation of a solute into a solvent.

solute The thing that gets dissolved in a solution.

solution *See* homogeneous mixture.

solvent The major component that dissolves a solute. Solvents are usually liquids, but can also be solids.

specific heat (C_p) The amount of energy required to heat one gram of a substance by one Kelvin at constant pressure.

spin quantum number Denoted by m_s, it distinguishes between the two electrons in an orbital. Possible values are +½ or –½.

spontaneous A process that takes place without any outside intervention.

standard conditions For gases, 1 atm and 273 K; for liquids, 1 M and 273 K.

standard temperature and pressure (STP) 0° C (273 K) and 1 atm.

stereoisomers Isomers that differ in three-dimensional structure from one another.

stoichiometry The method we use to relate the masses or volumes of the reactants and products of a chemical reaction to each other.

structural isomers Compounds having the same formula but differing in functional group or bonding pattern.

sublimation The process by which a solid becomes a gas without first becoming a liquid.

supercritical fluid A material at high enough conditions of temperature and pressure that it's no longer clear whether it's a gas or liquid.

supersaturated solution A solution that has dissolved more than the normal maximum possible amount of solute.

synthesis reaction When small molecules combine to form larger ones.

thermodynamics The study of free energy, enthalpy, and entropy.

titration The use of neutralization reactions to determine the concentration of an acid or base.

transition state The highest energy state between products and reactants in a chemical reaction.

triple point The conditions of temperature and pressure at which the liquid, gas, and solid phases of a material are all stable.

unit cell The smallest unit that can be stacked together to recreate a crystal.

unsaturated hydrocarbon A hydrocarbon containing at least one multiple bond.

unsaturated solution A solution that hasn't yet dissolved the maximum possible quantity of solute.

valence electrons The number of s- and p-electrons beyond the most recent noble gas.

valence shell electron pair repulsion theory (VSEPR) The shapes of covalent molecules depend on the fact that pairs of valence electrons tend to repel each other.

vapor pressure The vapor pressure of a liquid is the gas pressure in a closed container due to the molecules that have evaporated from the liquid.

voltaic cell Fancy word for "battery."

Index

A

B

C

H

I–J

K

L

P

Q–R

S

T

U

V

W-X-Y-Z